CORD ALGEBRA 1

PART A 1

Mathematics in Context

Developed by the
Center for Occupational Research and Development
Waco, Texas

JOIN US ON THE INTERNET
WWW: http://www.thomson.com
EMAIL: findit@kiosk.thomson.com   A service of I(T)P®

South-Western Educational Publishing
an International Thomson Publishing company I(T)P®

Cincinnati • Albany, NY • Belmont, CA • Bonn • Boston • Detroit • Johannesburg • London • Madrid
Melbourne • Mexico City • New York • Paris • Singapore • Tokyo • Toronto • Washington

CHAPTER 3

Using Formulas

CHAPTER 4

Solving Linear Equations

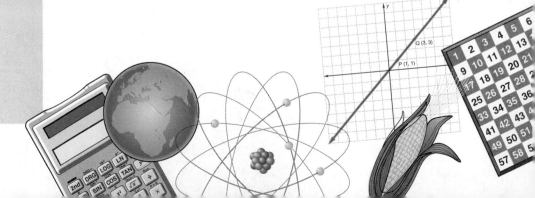

CHAPTER 5

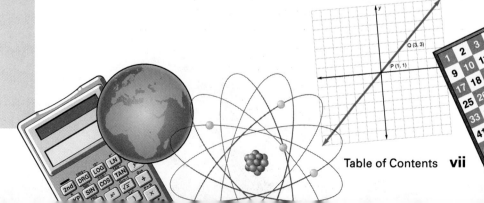

Graphing Linear Functions

CHAPTER 6

Nonlinear Functions

Also

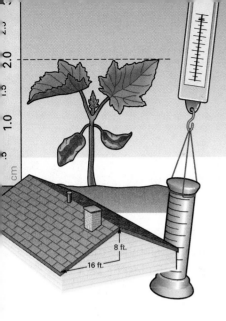

ALGEBRA AT WORK

"Why do I have to take an algebra course?" This is a great question and deserves an answer. Algebra is one of the cornerstones of today's technology boom. Without algebra there would not exist a single TV, radio, telephone, microwave oven, or other gadget that makes modern life so comfortable and interesting. Once you learn to use algebra, you will have the opportunity to help design and create the next generation of technology. You will secure yourself a place in tomorrow's workforce. But since all of this may seem mind boggling and too far in the future, here are just three of the ways algebra is used at work today.

USING ALGEBRA

Chen works for an aerospace company designing solar panels for satellites. His company is upgrading its environmental monitoring satellites by adding several new cameras to each. This upgrade is expensive. To reduce costs, the company wants to continue using the same type of solar panel that is already on the satellites. Chen is asked to find if the solar panels can handle the additional power requirements of the upgraded satellite. The power output of a solar panel depends on its area. The solar panels are rectangular in shape and are 3 feet wide by 5 feet high. To find the area of a solar panel, Chen does the following calculation:

$$\text{Solar panel area} = 3 \text{ ft} \times 5 \text{ ft} = 15 \text{ square feet}$$

Chen uses this area to find the total power output of the solar panels. After completing his calculations, Chen reports that the solar panels will power the upgraded satellite.

John is in charge of environmental safety at a large oil refinery. At 10 AM one morning, he received notice that one of the storage tanks on an oil tanker had begun leaking. John immediately deployed a cleanup crew. The cleanup crew reported back at 10:20 AM. The oil spill had already spread 200 feet to the north of the tanker. John knew he had a responsibility to report both the direction of the spill and its speed to the proper authorities. To find the speed of the oil spill, John does the following calculation:

$$\text{Oil spill speed} = \frac{200 \text{ ft}}{20 \text{ min}} = 10 \frac{\text{ft}}{\text{min}}$$

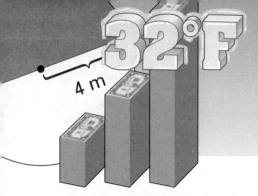

ALGEBRA AT WORK

John reported that the spill was advancing at a rate of $10\,\dfrac{\text{ft}}{\text{min}}$ in a northerly direction.

Tracy works for a data processing company. She is responsible for maintaining the company's computer system. The operating manual states that the room containing the computer system must be kept at a temperature of less than 20°C. The thermostat in the computer room reads 74°F. To determine if she needs to take any action, Tracy calculates the following:

$$\text{Temperature }°C = (74 - 32) \times \frac{5}{9} = 23.3°C$$

Tracy takes immediate action to lower the computer room temperature.

Because Chen, John, and Tracy understand how to use algebra, they are able to solve problems that affect their jobs. Each computation they do is based on a mathematical expression called a **formula**. Chen, John, and Tracy's job performance is directly related to how skillfully they use these formulas to solve problems.

To succeed in the workplace, you must be able to use problem solving skills. *CORD Algebra 1* is designed to provide you with these skills. Each chapter contains a number of exercises that relate directly to workplace problems. As you learn algebra, you will use it to solve problems involving agriculture, business and marketing, family and consumer science, health, and industrial technology. You will have a chance to sharpen your problem solving skills further by completing laboratory activities. These activities will test your knowledge of algebra and introduce you to important workplace skills, such as measuring, data collection, and data analysis. *CORD Algebra 1* will help you learn mathematics by using both your mind and your hands.

What is Algebra?

Algebra is a branch of mathematics that enables you to work with symbols instead of numbers. Take another look at the three

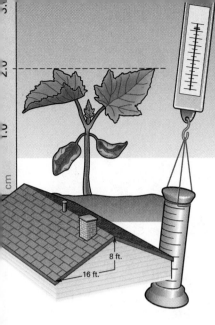

ALGEBRA AT WORK

situations involving Chen, John, and Tracy. They all used numbers. How did Chen, John, and Tracy know how to associate the numbers in the problems—such as length, height, time, speed, and temperature—with each other? In Chen's case, he used the formula for the area of a rectangle. John used a rate formula, and Tracy used a temperature conversion formula. Algebra provides tools for relating numbers, symbols, and formulas. Here is how it works.

Before Chen could begin multiplying numbers, he had to know how to calculate the area of a rectangle. The formula for the area of a rectangle is

$$A = lw$$

In this formula, the symbol A represents the area of a rectangle with length l and width (or height) w. The symbols A, l, and w are called **variables**. A, l, and w are variable because they can represent many different values of area, length, or width. A mathematical expression that contains a variable is called an **algebraic expression**. Examine Chen's calculations:

Solar panel area = 3 ft $\times$ 5 ft = 15 square feet

$$A = l \times w$$

As you can see, Chen's calculation is based on the area formula. He simply replaced the variables l and w with numerical values that represent the solar panel's dimensions.

John used a rate formula to find the speed of the oil spill.

$$v = \frac{d}{t}$$

In this formula, t represents the time it takes for the oil to spread a certain distance, d represents the distance the oil moves, and v is the speed of the oil spill.

Complete the activities that follow. Then you will see how solving problems in the real world depends on understanding how algebra works.

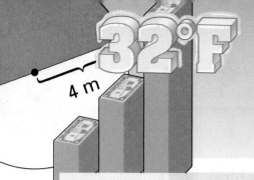

ALGEBRA AT WORK

ACTIVITY 1   The Rate Formula

In John's calculation, identify the numbers that stand for the variables t, d, and v.

$$\text{Oil spill speed} = \frac{200 \text{ ft}}{20 \text{ min}} = \frac{10 \text{ ft}}{\text{min}}$$

Complete the following table using the rate formula.

v	d	t
$200 \dfrac{\text{miles}}{\text{hours}}$	400 miles	2 hours
?	186,000 miles	1 second
?	12 meters	2 seconds
?	650 feet	50 seconds

Which speed in the table is closest to the speed of light?

Tracy used a temperature conversion formula that converts Fahrenheit temperature to Celsius temperature.

$$T_C = (T_F - 32) \times \frac{5}{9}$$

In this formula, the variable T_C represents a Celsius temperature and T_F represents a Fahrenheit temperature.

ACTIVITY 2   The Celsius Formula

Use Tracy's temperature conversion formula to convert 212°F to a Celsius temperature.

As you have just seen, algebra and its use of symbols gives us the means to write formulas. In turn, these formulas give us directions on how to calculate important quantities such as area, speed, and temperature. Algebra also provides a way for rearranging formulas to solve problems.

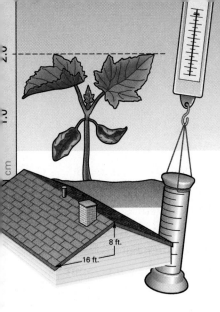

ALGEBRA AT WORK

Chen's company discovered a problem with the new satellite. The weight and location of the added cameras make the satellite hard to control and difficult to place in the proper orbit. One solution to this problem is to reduce the length of the solar panels to 3 feet. Chen knows the area of the solar panels has to stay at 15 square feet; otherwise the satellite will not have enough power to operate. This means the width of the solar panel has to increase. To find this new width, Chen rearranges the formula.

$$w = \frac{A}{l}$$

This new formula can be called the *width formula*.

ACTIVITY 3   The Width Formula

Calculate the new width of the satellite solar panel.

Once John reports the oil spill to the harbor authorities, he is faced with the challenge of containing the spill. John has no containment equipment at his refinery. The closest storage facility is 2 hours away. John knows it will take at least 4 hours to receive the equipment after he makes initial contact with the storage facility. To order the correct amount of equipment, John needs to know the approximate distance the oil spill will spread in 4 hours. John uses algebra to help him rewrite the rate formula in the following form:

$$d = vt$$

Now, the formula meets the need for calculating the distance, d, that the oil spill will spread after time, t.

ACTIVITY 4   The Distance Formula

How far will the oil spread in 4 hours?

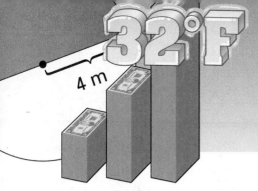

ALGEBRA AT WORK

Tracy has just received a new piece of equipment to install in her company's computer system. The instructions for operating this equipment state that it must be operated at a room temperature of less than 15°C. To what Fahrenheit temperature should Tracy set the room thermostat to meet the new equipment requirement? To answer this question, Tracy uses algebra to rewrite the temperature conversion formula.

$$T_F = \frac{9}{5} T_C + 32$$

ACTIVITY 5   The Fahrenheit Formula

Calculate the new temperature setting for the computer room thermostat.

Functions

You have seen that algebra can be used to write formulas. You have also found that algebra allows you to rewrite formulas. But the real power of algebra lies in how it is used to create formulas. Recall the formula for the area of a rectangle.

$$A = lw$$

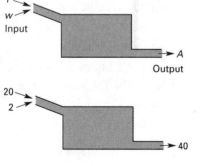

You can think of this formula as an input-output machine. If you input a value for l and w, the output is the area of a rectangle.

This formula also shows us that the output of the machine is dependent on the input. In the case of the area formula, the area of a rectangle depends on its length and width. Does this make sense? To answer this question, complete the following tables:

Table 1	l	w	A
	20 feet	2 feet	?
	20 feet	6 feet	?
	20 feet	10 feet	?

ALGEBRA AT WORK

Table 2

l	w	A
20 feet	2 feet	?
30 feet	2 feet	?
40 feet	2 feet	?

In Table 1, the length l remains fixed, and the width w is varied. As w varies, so does the area of the rectangle, A. Thus, the area of a rectangle does depend on its width. Table 2 demonstrates that the area of a rectangle also depends on its length. Because of the dependence of A on both l and w, the variable A is called a **dependent variable**. The variables l and w are called **independent variables**. Dependent variables are the outputs of our input-output machine, and independent variables are the inputs.

ACTIVITY 6   The Rate Formula

Complete the following tables for the rate formula. Then identify the dependent and independent variables.

Table 3

d	t	v
30 feet	2 seconds	?
30 feet	10 seconds	?
30 feet	15 seconds	?

Table 4

d	t	v
30 feet	2 seconds	?
60 feet	2 seconds	?
90 feet	2 seconds	?

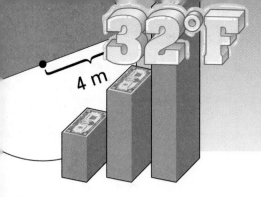

ALGEBRA AT WORK

The relationship between the area of a rectangle and its length and width is called a **function**. A function is a mathematical way of describing certain relationships between quantities. The relationship between speed, distance, and time in the rate formula also is a function. The relationship between Celsius and Fahrenheit temperatures in a temperature conversion formula is a function as well. Starting in Chapter 5 and continuing for the rest of your work in algebra, functions will become a central part of your study. Here are just two examples of how functions have changed our technology today.

Think about diving into a lake. What happens to your ears as you go deeper into the water? You feel more pressure. Thus, there is a relationship between the pressure water exerts on your ears and the height of water above you. This relationship is a function. Scientists discovered this relationship back in the 1700s. From this relationship, scientists have found formulas that help design today's fastest and most maneuverable jet airplanes.

Turn on a radio. Stand a certain distance from the radio and then slowly begin to move away from it. What happens to the loudness of the sound? It decreases. The loudness of the sound is related to the distance from the source of the sound. This relationship is a function that scientists can express as a formula. This same formula is used extensively in designing today's high resolution stereo speakers.

Algebra and Your Future

The future is full of exciting technological challenges—faster planes and more powerful stereo systems. If you want to be a part of this future, you must be able to use the concepts of algebra. As the previous examples have shown, algebra is the language of technology. Becoming fluent in this language will help secure your future success. So why delay the future? Now is the time to begin your study of algebra.

CORD ALGEBRA 1

PART
A 1

Mathematics in Context

CHAPTER 1

WHY SHOULD I LEARN THIS?

Many of the tools of modern technology—computer chips, lasers, and high resolution TV—were invented by people skilled in the use of numbers and vectors. As technology becomes a bigger part of our everyday lives, the need for these skills is growing. Build the skills for your future now by exploring the concepts in this chapter.

INTEGERS AND VECTORS

OBJECTIVES

1. Identify integers.
2. Find the absolute value of an integer.
3. Add, subtract, multiply, and divide integers.
4. Find the magnitude and direction of a vector.
5. Solve problems involving integers and vectors.

Numbers such as 3 are used to model many everyday situations.

1. A city is 3 feet below sea level. 2. A stock index drops 3 points.

3. You walk north 3 miles.   4. A friend walks south 3 miles.

5. A store takes in 3 dollars more than it pays out.

To show that you walked north 3 miles, you can use the positive number, $+3$. To show that your friend walked 3 miles south, you can use the negative number, -3. The number 3 shows how far each of you walked. The positive and negative signs show which direction.

Directed line segments or *vectors* represented by arrows also model many situations. In each of the vector drawings, a specific direction and magnitude is represented.

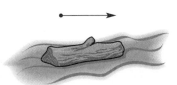

River Current 3 mph, east

Force on Rope 3 lb, 45°N of E

Notice each vector represents some actual physical quantity such as a river current or a pulling force.

This chapter will show you how to use positive and negative numbers and how to draw and use vectors to solve problems. As you view the video for this chapter, notice how people on the job use integers and vectors.

LESSON 1.1 IDENTIFYING INTEGERS

A temperature rise of 5 degrees can be represented by the symbol +5 and read, "positive five." A temperature drop of 3 degrees can be represented by the symbol −3 and read "negative three."

Positive and Negative Numbers

A number preceded by a positive sign (+) is a positive number. A number preceded by a minus sign (−) is a negative number. If a number is not preceded by either sign, it is considered positive.

Critical Thinking   How is placing a minus sign in front of a number different than placing it between two numbers?

A **number line** is used to picture positive and negative numbers. Follow the steps in the following activities to draw a number line.

ACTIVITY 1   Constructing a Number Line

1 In the center of a sheet of paper, draw a point and label it 0. This zero point is the center of your number line and is called the **origin.**

$$\bullet$$
$$0$$

2 Start at the origin and draw a horizontal ray to the right. Show that this part of the number line continues forever by drawing an arrow pointing to the right.

3 Mark a point on the ray about a thumb's width to the right of the origin. Label this point 1. The distance from 0 to 1 on the number line is called the **unit** for the number line. Once you choose the unit length, it must remain the same for the entire line.

4 Now start at 1. Use a compass or piece of string to mark off one unit to the right of 1. Label this mark 2.

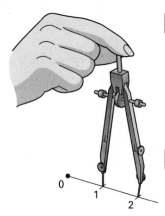

5 Now continue to mark units until you reach the arrowhead. Keep the distance between marks equal to 1 unit. Remember to label each mark.

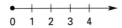

The marks you labeled to the right of zero represent **positive integers.** One way to describe the set of positive integers is to use three dots.

$$1, 2, 3 \ldots$$

The three dots indicate that the numbers continue in the same way. Notice that there is no largest positive integer.

Now continue your number line to represent the negative integers.

ACTIVITY 2   **Negative Numbers on the Number Line**

1 Start at the origin and extend the ray to the left. Draw an arrow pointing to the left to show that it goes forever to the left of 0.

2 Mark equal units to the left of 0. What point is one unit to the left of 0? Complete your number line. These numbers are called **negative integers.**

3 How many negative integers do you think there are? Is there a smallest negative integer?

Your number line should look similar to this one. The marks on the number line represent the set of all integers.

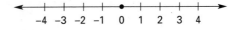

The Set of Integers
    The set of integers includes all the positive integers, all the negative integers, and zero.

At what two times did the radar equipment come closest to automatic shutdown? If you were adjusting the air conditioning, when would you want it to operate most effectively?

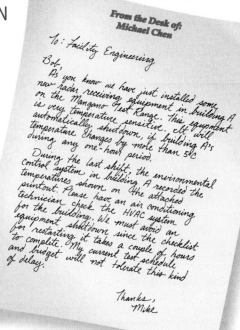

From the Desk of:
Michael Chen

To: Facility Engineering

Bob,

As you know we have just installed some new radar receiving equipment in building A on the Mangano Test Range. This equipment is very temperature sensitive. It will automatically shutdown, if building A's temperature changes by more than 5°C during any one-hour period.

During the last shift, the environmental control system in building A recorded the temperatures shown on the attached printout. Please have an air conditioning technician check the HVAC system for the building. We must avoid an equipment shutdown since the checklist for restarting it takes a couple of hours to complete. My current test schedule and budget will not tolerate this kind of delay.

Thanks,
Mike

ENVIRONMENTAL CONTROL DATA FOR			
Time	Temperature, C	Temperature Change, C	Relati Hum
8am	22		
9am	26	4	
10am	25	−1	
11am	26	1	
12	28	2	
1pm	28	0	
2pm	27	−1	
3pm	28	1	
4pm	25	−3	
5pm	24	−1	

LESSON ASSESSMENT

Think and Discuss

1 What is an integer?

2 Explain how a minus sign can be used in two ways.

3 Give two examples where negative integers model workplace situations.

4 How is a meter stick like a number line?

5 Explain how a number line is used to model the countdown of a space shuttle launch.

Practice and Problem Solving

Write an integer to model each situation.

6. An elevator goes down 15 floors.

7. A construction detour causes a 25 minute longer drive to work after school.

8. City Floral lost $18,500 the first year of business.

9. To land at the local airport, a plane must lose 2340 feet of altitude.

10. The stock market increased 22 points in the first hour of trading.

11. The deepest point in the Atlantic Ocean is 8648 meters below sea level.

12. A hot air balloon rose 515 feet.

13. Tim deposited $36 into his savings account.

14. Reynaldo gained 4 pounds during the first week of weight training.

15. The Smiths enlarged their patio by 400 square feet.

16. After a chemical reaction, the temperature of the metal dropped 52 degrees.

17. A mountain climber descends 1000 feet in one hour.

Mixed Review

Solve each problem.
You must decide whether to repair your car or replace it. However, you only have $2000 to spend. After checking several repair shops, you find it will cost $885 to repair the engine and $640 to repair the outside of the car. A good used car will cost $4800. You can get $1200 for trading in your old car.

18. What is the total cost to repair your old car?

19. After the trade-in, how much will you pay to buy a used car?

20. Given the information you now have, which choice would you make? Explain why.

LESSON 1.2 ABSOLUTE VALUE

The Number Line

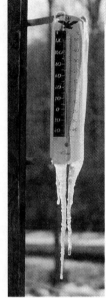

In Lesson 1.1, you used a number line to represent the set of integers. Numbers to the right of zero are positive. Numbers to the left of zero are negative. Zero is neither positive nor negative. A thermometer held horizontally is a good model for the integers. Most thermometers show two number lines with different scales. The scale is the size of one unit—the distance between 0 and 1.

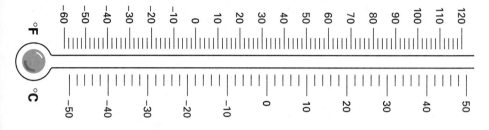

ACTIVITY 1   Using an Integer Model

Use either the °C or °F scale to answer questions 1–7 on your paper.

1 On which side of zero are the positive values?

2 On which side of zero are the negative values?

3 Which temperature is farther to the right: 20° or 30°?

4 Which temperature is higher (hotter): 20° or 30°?

5 Which temperature is farther to the right: −20° or −30°?

6 Which temperature is higher (hotter): −20° or −30°?

7 Which temperature is farther to the right: 20° or −30°?

Critical Thinking   If two numbers are on a number line, how can you determine which one is larger? Do any of the answers change if you use a different scale?

Think about the comparison of −20° and −30°. On a number line, −20 is to the right of −30. Thus, −20 is greater than −30. It is also true that −20° is a higher temperature than −30°. As you move to the *right* on the number line, the numbers *increase*.

Ongoing Assessment

Use a number line to determine which number is greater.

a. 4 or −6   b. −3 or −7   c. −4 or −2

The sentence *−20 is greater than −30* can be written

$$-20 > -30$$

The symbol $>$ is read "is greater than."

The sentence *−30 is less than −20* can be written

$$-30 < -20$$

The symbol $<$ is read "is less than."

Notice that the point of the less than or greater than symbol always points to the number of lesser value.

Absolute Value

Notice that −20 is 20 units from zero on the number line. The distance from zero to a number is called the **absolute value** of the number. The absolute value of −20 is 20. The symbol for absolute value is $|\ |$. Thus,

$$|-20| = 20$$

To remember the meaning of the absolute value symbol, read the first vertical line as "the distance of" and the second vertical line as "from zero."

Since −30 is 30 units from zero,

$$|-30| = 30$$

How far from zero is $+30$? Notice that $+30$ is exactly the same distance from zero as -30. Both numbers are 30 units from zero. Therefore, $+30$ and -30 have the same absolute value.

$$|+30| = |-30| = 30$$

Equal Absolute Values

If two numbers are the same distance from zero on the number line, the numbers have the same absolute value.

Ongoing Assessment

Use $<$, $>$ or $=$ to compare the absolute values.

a. $|-10|$ and $|-4|$ b. $|-3|$ and $|-5|$ c. $|-9|$ and $|9|$

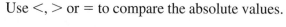

Here is one way to think of the relationship between -20 and -30.

-20 degrees Fahrenheit is hotter than -30 degrees Fahrenheit.

$$-20°F > -30°F$$

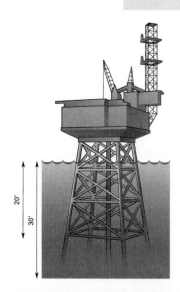

But 30 feet underwater (-30) is a greater depth than 20 feet underwater (-20).

$$|-30| > |-20|$$

In other words, -30 is farther from zero (the water surface) than -20.

LESSON ASSESSMENT

Think and Discuss

1 How can you use a thermometer to model the integers?

2 Compare the distance from the zero point on a thermometer for two temperatures that have the same absolute value.

3 Describe the relative location on a thermometer for one temperature that is greater than another.

4 What is the distance between two temperatures on a thermometer if the absolute values of the two temperatures are equal?

5 Why is $|-10| > |-5|$?

Compare the integers. Use < or >.

6. −3; 4 **7.** −6; −8 **8.** 5; −5 **9.** −4; −5

10. −1; 0 **11.** −20; −15 **12.** 12; 1 **13.** 0; −9

14. A time line is an example of a number line. Draw a time line and show where each event should be placed.

   A. Buddha lived in China around 500 BCE.

   B. The Old Egyptian Empire was just beginning around 2700 BCE.

   C. The great Epics of India were being written around 1200 BCE.

   D. Hammurabi was the ruler of the Tigris-Euphrates Valley around 1800 BCE.

   E. The Hebrews left Egypt around 1500 BCE.

   F. The Phoenicians traded throughout the Mediterranean around 1000 BCE.

Evaluate each expression.

15. $|-7|$ **16.** $|10|$

17. $|0|$ **18.** $|-23|$

19. $|-6|$ **20.** $|-18|$

21. $-|-25|$ **22.** $|35-17|$

Each employee production team at the Captain Tire Company has a goal of producing 278 tires per day. Each day, your team's production is recorded as a deviation from the 278 tires. If you produce 281 tires, the deviation is +3. If you produce 277 tires, the deviation is −1. On Monday, you received this part of an e-mail message:

> Daily bonuses will be paid to production teams only when your team's product deviation is greater than −5

23. Will you make a bonus if your team produces 270 tires for the day?

24. How many tires does your team have to produce to make a bonus?

Mixed Review

25. Rashid's take-home pay is $2045 each month. Rashid has completed the budget shown below. Find the amount Rashid will have left after he pays his monthly expenses.

Rent	$475	Utilities	$125
Phone	$ 46	Cable	$ 52
Car payment	$279	Car expense	$ 85
Food	$418	Charge cards	$ 78

LESSON 1.3 ADDING INTEGERS

You can use a number line to model integer addition. Adding a positive number results in a move to the right. Adding a negative number results in a move to the left.

Adding Integers with the Same Sign

The temperature is 12°F at noon. Suddenly the temperature increases by 3°. What is the current temperature?

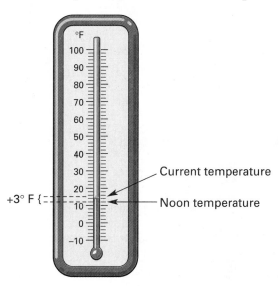

+3° F {

Current temperature

Noon temperature

The thermometer shows the new temperature is 15°. This is an example of adding two integers with the same sign. In this case, both integers are positive. Thus,

$$+12 + (+3) = +15 \text{ or } 12 + 3 = 15$$

Adding two positive integers is just like adding two numbers in arithmetic. What happens when you add two negative integers?

If a football team loses four yards on the first down, the loss is represented by -4. If they lose another three yards (-3) on the next down, they have lost seven yards (-7) in all. A mathematical sentence is used to represent this situation.

$$-4 + (-3) = -7$$

In this sentence, the $+$ sign is used as an arithmetic operation. This sign tells you to add two negative numbers. The second negative number is enclosed in parentheses to avoid confusing the + symbol for addition with the symbol that shows the sign of the number.

To model the operation $-4 + (-3) = -7$ on a number line, begin at zero. Draw an arrow called a **vector** four units to the left to represent -4. From (-4), draw another vector 3 units to the left to represent -3. Notice that the vectors are defined both by magnitude (the length of the arrow) and direction. At what point is the tip of the second arrow?

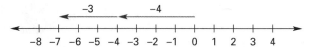

This model shows that

$$-4 + (-3) = -7$$

ACTIVITY 1   Adding Negative Integers

1 Choose 5 pairs of negative numbers.

2 Use a number line to add each pair.

3 Write a mathematical sentence to model each addition.

4 What is the sign of each sum?

When two or more negative numbers are added together, the result is always a negative number. This is exactly the same rule that you use with positive numbers: When two or more positive numbers are added together, the result is always a positive number.

> **Adding Integers with the Same Sign**
> When you add two or more integers with the same sign,
>
>   1. add the absolute values,
> and
>   2. write the total with the sign of the integers.

A calculator can be used to add negative integers. First, add -328 and -49 on your calculator. Then compare your method to the one shown on the opposite page.

LESSON 1.3 ADDING INTEGERS

You can use a number line to model integer addition. Adding a positive number results in a move to the right. Adding a negative number results in a move to the left.

Adding Integers with the Same Sign

The temperature is 12°F at noon. Suddenly the temperature increases by 3°. What is the current temperature?

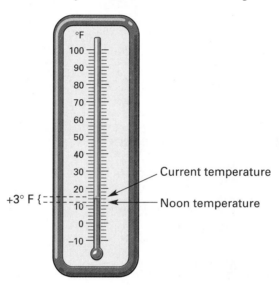

The thermometer shows the new temperature is 15°. This is an example of adding two integers with the same sign. In this case, both integers are positive. Thus,

$$+12 + (+3) = +15 \text{ or } 12 + 3 = 15$$

Adding two positive integers is just like adding two numbers in arithmetic. What happens when you add two negative integers?

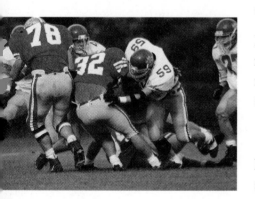

If a football team loses four yards on the first down, the loss is represented by -4. If they lose another three yards (-3) on the next down, they have lost seven yards (-7) in all. A mathematical sentence is used to represent this situation.

$$-4 + (-3) = -7$$

In this sentence, the $+$ sign is used as an arithmetic operation. This sign tells you to add two negative numbers. The second negative number is enclosed in parentheses to avoid confusing the $+$ symbol for addition with the symbol that shows the sign of the number.

To model the operation $-4 + (-3) = -7$ on a number line, begin at zero. Draw an arrow called a **vector** four units to the left to represent -4. From (-4), draw another vector 3 units to the left to represent -3. Notice that the vectors are defined both by magnitude (the length of the arrow) and direction. At what point is the tip of the second arrow?

This model shows that

$$-4 + (-3) = -7$$

ACTIVITY 1   Adding Negative Integers

1 Choose 5 pairs of negative numbers.

2 Use a number line to add each pair.

3 Write a mathematical sentence to model each addition.

4 What is the sign of each sum?

When two or more negative numbers are added together, the result is always a negative number. This is exactly the same rule that you use with positive numbers: When two or more positive numbers are added together, the result is always a positive number.

Adding Integers with the Same Sign
When you add two or more integers with the same sign,

　　　1. add the absolute values,
and
　　　2. write the total with the sign of the integers.

A calculator can be used to add negative integers. First, add -328 and -49 on your calculator. Then compare your method to the one shown on the opposite page.

Enter the number 328

Press the ⊞∕⊟ key. (Remember, the ⊞∕⊟ key changes the sign of the number in the display.)

Press ⊞

Enter the number 49

Press the ⊞∕⊟ key

Press ⊟

Did you get -377?

Critical Thinking You can also add the absolute values. First press the keys for $328 + 49$. Then, write the total with a negative sign in front. Explain why this works.

Adding Integers with Different Signs

How do you add two numbers that have different signs? Recall the football example. This time, let the first down result in a loss of 7 yards and the next down result in a gain of 4 yards. The final result of the two plays is a net loss of 3 yards. A mathematical sentence can be used to represent this situation.

$$-7 + 4 = -3$$

You can also model $-7 + 4 = -3$ on a number line. Draw a vector from zero to -7 to represent a loss of seven yards. Since the next play is a gain of 4 yards, draw a vector from -7 four units to the right. Where do you finish on the number line?

ACTIVITY 2   Adding Integers Using Absolute Value

1 Explain why the following sketch shows the sum of $-7 + 4$.

2 Notice that the vector representing -7 is longer than the vector representing 4. How does this help you determine the sign of the answer?

3 What is the difference between the absolute value of -7 and the absolute value of 4? What is the sign of the sum?

4 Explain how to use absolute value to find the sum of 3 and -8.

Did you notice that it is a lot easier to add two numbers with different signs than it is to explain it?

Adding Integers with Different Signs
When you add two integers with different signs

    1. subtract the absolute values,

and

    2. write the difference with the sign of the integer having the larger absolute value.

To add $-7 + 4$, first subtract the absolute values.

$$|-7| - |4| = 7 - 4 = 3$$

Since $|-7| > |4|$, write the difference as -3.

$$-7 + 4 = -3$$

Ongoing Assessment

Find each sum.

a. $-6 + -7$ **b.** $-2 + 5$ **c.** $9 + (-14)$

CULTURAL CONNECTION

For a very long time, people have used the idea of negative numbers. Among the first to work with both positive and negative numbers were the Chinese. They did so by using colored rods. Any two colors will work, as long as they are easily contrasted. Here is how a Chinese merchant might have calculated his finances by using red rods to represent negative integers and blue rods to represent positive integers.

The merchant might show a loss of 6 followed by a gain of 4 by placing 6 red rods and 4 blue rods in a container.

Negative 6

Positive 4

The merchant would then remove 4 red rods and 4 blue rods. Why?

Since there are 2 red rods left in the container, the net result is a loss of 2.

Use the colored rod model to explain the rule for adding integers with different signs. Find the difference of the number of rods. Use the sign associated with the greater number of rods.

LESSON ASSESSMENT

1 Explain how to use a number line to add two positive integers.

2 Explain how to use a number line to add two negative integers.

3 Explain how to use a number line to add a positive and a negative integer.

4 How is absolute value used to add two integers with the same sign?

5 How is absolute value used to add two integers with different signs?

Practice and Problem Solving

Add.

6. $-3 + 5$ **7.** $-8 + 8$ **8.** $6 + (-4)$

9. $-9 + (-6)$ **10.** $-23 + 12$ **11.** $-14 + (-11)$

12. $13 + (-7)$ **13.** $-16 + (-18)$ **14.** $-21 + 15$

15. $-23 + (-9)$ **16.** $-9 + (-23)$ **17.** $27 + (-27)$

18. $45 + (-26)$ **19.** $-63 + (-37)$ **20.** $-59 + (-67)$

21. During one week of testing a new generator, electrical output dropped 328 kilowatts. During the next three weeks, output increased 143 kilowatts, decreased 37 kilowatts, and increased 219 kilowatts. Use integers to describe the net decrease or increase.

A proton has a charge of positive one. An electron has a charge of negative one. Find the total charge of an ion with

22. 12 protons and 15 electrons.

23. 21 protons and 19 electrons.

24. Mark the following amounts on a number line.

    a. gain of $48   **b.** loss of $36   **c.** loss of $12
    d. loss of $18   **e.** gain of $27   **f.** loss of $27

The chart shows several altitudes above or below sea level.

Caspian Sea	−28 meters
Death Valley	−86 meters
Mount Everest	+8848 meters
Mount McAuthur	+4344 meters
Pacific Ocean	−10,912 meters
Sea Level	0 meters

Compare the altitudes of each pair by writing an inequality.

25. Death Valley; Caspian Sea

26. Mount Everest; Mount McAuthur

27. Sea Level; Death Valley

28. Mount Everest; Pacific Ocean

29. Mount McAuthur; Sea Level

30. List all six altitudes in order from least to greatest.

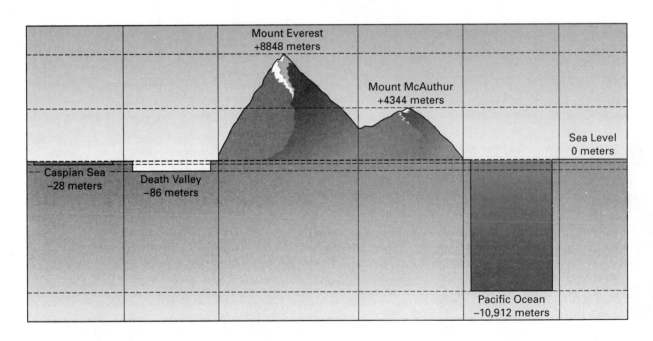

LESSON 1.4 SUBTRACTING INTEGERS

The number line is also a good model for integer subtraction. To model $8 - (-4)$, think of -4 in another way.

Opposites

Notice the position of -4 and 4 on the number line. Although they are the same distance from zero, they are on **opposite** sides of zero. From the number line, you can see that

-4 is *"opposite of 4"* and 4 is *"opposite of -4."*

> **The Opposite of an Integer**
> Two integers are opposites if they are
>
> 1. on opposite sides of zero,
>
> and
>
> 2. the same distance from zero.

Thus,

$$\text{the opposite of } (-4) = 4$$

$$\text{and the opposite of } 4 = -4$$

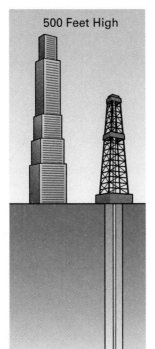

500 Feet High

−500 Feet Deep

A minus sign $(-)$ is used to indicate "the opposite of" a number. The opposite of -4 can also be written as $-(-4)$. Thus, "the opposite of negative four is positive four" can be written as follows:

$$\text{opposite sign} \rightarrow -(-4) = 4$$
$$\uparrow$$
$$\text{negative sign}$$

Some calculators have a ⊞⊟ key. If your calculator has one, experiment with it. How is this key used to find the opposite of an integer? How is the ⊞⊟ key used to find the absolute value of an integer?

Adding Opposites

ACTIVITY 1   The Sum of Two Opposites

1. Choose five pairs of opposite numbers.

2. Find the sum of each pair.

3. What is the sum of each pair of opposites?

> **Addition of Opposites**
> The sum of two opposites is always zero.

Opposites are used to find the difference between two integers.

ACTIVITY 2   Use a Calculator to Subtract Integers

1. Complete each set of directions using a scientific calculator.

   a. $-10 - (-6)$

   Enter 10.
   Press the ⌗ key.
   Press the ⊖ key.
   Enter 6.
   Press the ⌗ key.
   Press the ⊜ key.

   b. $-10 + 6$

   Enter 10.
   Press the ⌗ key.
   Press the ⊕ key.
   Enter 6.
   Press the ⊜ key.

2. What is the result in each case? Why are the results equal?

3. Which method do you prefer? Explain why.

4. How are the expressions $-10 - (-6)$ and $-10 + 6$ the same? How are they different?

5. Explain how you can use addition to solve a subtraction problem.

The calculator problem is an example of the rule for subtracting a negative integer. However, the rule also works when you subtract a positive integer.

Subtracting Integers

You can subtract an integer by adding its opposite.

Both adding and subtracting occur often in the workplace. For instance, the wings of an airplane are attached to the airplane fuselage using rivets. After the rivets are in place, they are cured by heating them to 125°C and holding them at this temperature for a specified period of time. This curing process insures that the joint between the wings and the fuselage has maximum strength.

During the curing process, a technician monitors the temperature and records hourly readings.

Time	Deviation from 125°C
8 AM	+1
9 AM	+5
10 AM	−4
11 AM	+2
12 AM	0

Since the temperature is supposed to be 125°C, +1 indicates a temperature of 126°C and −3 indicates a temperature of 122°C.

Every four hours, the technician is required to report the hourly changes in the temperature to the engineer in charge of the plane's assembly.

To find the hourly change between 8 AM and 9 AM, the technician subtracts a positive number.

$$(+5) - (+1) = 5 - 1 = 4$$

Thus, during this hour, the temperature increased 4° from 126°C to 130°C. How can you find the temperature change between 9 AM and 10 AM?

$$(-4) - (+5) = ?$$

Since you can subtract an integer by adding its opposite, change +5 to −5 and add −5 to −4. Mathematically, this statement is modeled as

$$(-4) + (-5) = -9$$

From 9 AM to 10 AM the temperature decreased by 9° from 130°C to 121°C.

The temperature change from 10 AM to 11 AM is

$$(+2) - (-4) = (+2) + (+4) = +6$$

From 10 AM to 11 AM, the temperature increased by 6° from 121°C to 127°C. What was the temperature change from 11 AM to 12 AM?

Ongoing Assessment

Solve each problem without paper and pencil.

1. $9 + 6$ **2.** $9 + (-6)$ **3.** $9 - 6$ **4.** $9 - (-6)$

5. $-9 + 6$ **6.** $-9 + (-6)$ **7.** $-9 - 6$ **8.** $-9 - (-6)$

LESSON ASSESSMENT

Think and Discuss

1 Explain how to subtract an integer by using its opposite.

2 You buy a $15 CD at a music store. Use a number line to show that adding $15 to the account balance of the music store is the opposite of subtracting $15 from your checking account balance.

Practice and Problem Solving

Subtract.

3. $9 - (-5)$ **4.** $8 - 11$

5. $-7 - (-6)$ **6.** $-4 - 3$

7. $-12 - (-15)$ **8.** $-14 - 16$

9. $18 - (-14)$ **10.** $21 - (-19)$

Perform the indicated operations.

11. $-21 + (-15) - (-16)$ **12.** $38 - 45 + 26$

13. $-29 + 53 - 42$ **14.** $-31 - (-48) - (-16)$

15. $-56 + (56) - 18$ **16.** $75 - 96 + 43$

Tara made two deposits and two withdrawals from her bank account. She started with $781.08 in her account. Find the amount Tara has in her account after each transaction.

17. September 15; + $116.58   **18.** September 20; − $216.25

19. October 1; − $201.88   **20.** October 10; + $196.39

Mixed Review

The average rainfall for February is 4.21 inches. Following is the deviation from the average for five different years. Find the February rainfall for each year.

21. +1.85   **22.** −0.68   **23.** −1.36   **24.** −1.5   **25.** +0.7

The height of the Empire State Building is 1250 feet. Use integers to compare each of the following buildings with the Empire State Building.

26. Sears Tower; 1454 feet   **27.** John Hancock; 1127 feet

28. Amoco; 1136 feet   **29.** World Trade North; 1368 feet

World Trade
Center North,
1368 feet

Empire State
Building,
1250 feet

LESSON 1.5 MULTIPLYING AND DIVIDING INTEGERS

You can use a multiplication table to find the product of two positive integers such as 2×3.

×	0	1	2	3
3	0	3	6	9
2	0	2	4	6
1	0	1	2	3
0	0	0	0	0

This multiplication table has four rows and four columns. The row labels are at the left of the table. The column labels are at the top of the table. The numbers in the table are entries. The entry in row 2 and the column headed by 3 is 2×3. Since $2 \times 3 = 6$, the entry is 6.

Critical Thinking   Why are there so many zeros in the table?

Multiplying Negative Numbers

ACTIVITY 1   **Extending the Multiplication Table**

1   Copy this partial table.

×	-3	-2	-1	0	1	2	3
3				0	3	6	9
2				0	2	4	6
1				0	1	2	3
0	0	0	0	0	0	0	0
-1				0			
-2				0			
-3				0			

2   The table is made up of four parts, called **quadrants.** What are the signs of the entries in the top right quadrant? Explain how this table shows that the product of two positive integers is a positive integer.

3 Start at 9. Move across the first row of entries from right to left. Do the entries increase or decrease? By how much? Describe the pattern in each row as you move from right to left.

4 Use the pattern from step 3 to complete the entries in the top left quadrant of the table.

×	−3	−2	−1	0	1	2	3
3	?	?	?	0	3	6	9
2	?	?	?	0	2	4	6
1	?	?	?	0	1	2	3
0	0	0	0	0	0	0	0

5 Does your table look like the one below?

×	−3	−2	−1	0	1	2	3
3	−9	−6	−3	0	3	6	9
2	−6	−4	−2	0	2	4	6
1	−3	−2	−1	0	1	2	3
0	0	0	0	0	0	0	0

Explain how you can use the table to show that the product of a negative integer and a positive integer is a negative integer.

6 Extend the table downward so that it has rows numbered −1, −2, and −3. Use patterns to complete the lower half of the table. Describe the patterns you see in the table.

7 Does your table look like the one below?

×	−3	−2	−1	0	1	2	3
3	−9	−6	−3	0	3	6	9
2	−6	−4	−2	0	2	4	6
1	−3	−2	−1	0	1	2	3
0	0	0	0	0	0	0	0
−1	3	2	1	0	−1	−2	−3
−2	6	4	2	0	−2	−4	−6
−3	9	6	3	0	−3	−6	−9

Explain how you can use the table to show that the product of two negative integers is a positive integer.

Multiplying Integers

The product of two integers with the same sign is positive.

The product of two integers with different signs is negative.

Here is a short form of the rule for multiplying integers:

Same signs $\rightarrow$ **Positive product**

Different signs $\rightarrow$ **Negative product**

When using a calculator to find a product, first decide on the sign of the product. For example, when multiplying $128 \times (-78)$, look at the signs of the integers. In this case, you are multiplying two integers with different signs. The product will be negative.

Now multiply the absolute values of the numbers and write the product with the correct sign. Thus, $128 \times (-78) = -9984$.

You can also find the product $128 \times (-78)$ by using the $\boxed{+/-}$ key on a scientific calculator.

$128 \times (-78) = ?$

Enter 128
Press $\boxed{\times}$
Enter 78
Press $\boxed{+/-}$
Press $\boxed{=}$

Does the window display -9984?

Dividing Integers

Multiplying and dividing are **inverse operations**; that is, they undo each other. The rule for finding the sign of the quotient of two integers is exactly the same as for multiplying integers.

Dividing Integers

The quotient of two integers with the same sign is positive.

The quotient of two integers with different signs is negative.

Find each quotient mentally.

a. $12 \div 4$ **b.** $12 \div (-4)$ **c.** $-12 \div 4$ **d.** $-12 \div (-4)$

Now find $-13{,}650 \div 26$ using your calculator.

It's *still* not full! Bring more zeros!

Be sure to press the ⊞∕⊟ key after entering 13650. Did you get -525? Now try

$$-13{,}650 \div (-26.1)$$

The display should show 522.9885057. Rounded to the nearest whole number, the answer is 523.

Critical Thinking   Use your calculator to divide 5 by zero. What happens? Why is it impossible to divide by zero?

LESSON ASSESSMENT

Think and Discuss

1 How do you know which sign to use when you multiply two integers?

2 What are the steps for multiplying -24.5 by -6 on a calculator? What is the product?

3 What are the steps for dividing -7.5 by -0.5 on a calculator? What is the quotient?

4 Explain why 1.13 is not a reasonable quotient for -3.4 divided by 3.

5 When more than two integers are multiplied, how is the sign of the product chosen?

Practice and Problem Solving

Perform the indicated operations.

6. -3×-8 **7.** $-6 \div -3$ **8.** -4×7 **9.** $-18 \div 2$

10. 8×-7 **11.** $63 \div -9$ **12.** -12×-12 **13.** -24×3

14. 15×-6 **15.** $-25 \div -3$ **16.** $-8 \div -6$ **17.** $38 \div -8$

To find the average of 5 numbers, find the sum of the numbers and then divide that sum by 5.

18. Find the average of 8, 3, 7, 9, and 3.

19. Subtract each number in Exercise 18 from the average of the numbers. Write the differences using negative and positive integers.

20. Find the average of −16, 11, 12, −20, and 18.

21. Find the difference between each number in Exercise 20 and the average of the numbers. Write the differences using negative and positive integers.

Mixed Review

23. How much greater than −26 is −15? Explain your reasoning.

Use the table for Exercises 24−26.

COPIES	
1–99	$0.10
100–499	$0.09
500–999	$0.08
1000+	$0.07

24. How much does it cost to make 499 copies?

25. How much does it cost to make 500 copies?

26. How much does it cost to make 445 copies?

27. A customer said that if he had between 445 and 500 sheets to copy, he would rather pay for 500 sheets. Explain why.

In a previous lesson, you used **vectors** to add and subtract integers. A vector is modeled by an arrow that shows both magnitude and direction. The length of the arrow indicates the magnitude of the vector. The arrow indicates the direction of the vector. For example, the ship's velocity vector has magnitude 3 and direction to the right. This vector indicates that the ship's velocity is 3 miles per hour to the east.

3 miles/hr, east

Ongoing Assessment

Use a number line and draw a vector to show a move from 1 to 4. What is the length of the vector? What is the direction of the vector?

The arrow used to show the speed and direction of the ship could represent any vector with a magnitude of 3 and the same direction. For example, it could represent a force of three pounds pulling to the right. The vector could also represent a current flowing at three miles per hour downstream.

Force on Rope 3 lb, 45°N of E   River Current 3 mph, east

The way the arrow points tells you the direction of the vector. The length of the arrow tells you the magnitude or size. To determine magnitude, you must have a scale such as a number line or a number written next to the arrow.

Vectors can show the position of an object after two or more different moves.

Suppose a ship begins at point *A* and sails 10 miles north to point *B*. It then turns and sails 10 miles east to point *C*. Vectors *AB* and *BC* describe the two parts of the ship's journey. Vector *AB* has a magnitude of 10 miles and a direction of north. Vector *BC* also has a magnitude of 10 miles, but the direction is east. Why are vectors *AB* and *BC* *not* equal?

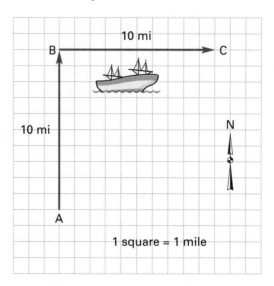

Vectors can model the addition and subtraction of several integers. Remember that subtracting an integer is the same as adding its opposite. The **opposite of a vector** is a vector that is the same size, but drawn in the opposite direction.

ACTIVITY   **Opposite Vectors**

1 Construct a number line.

2 Start at the origin and draw a vector to the right 4 units in magnitude.

3 Start at the origin and draw a vector to the left 4 units in magnitude.

4 Explain why these two vectors are called opposites.

Critical Thinking   Draw a vector representing a ship's movement 3 miles to the east. Then draw a vector to represent the opposite of the ship's vector. What is the result of a ship's journey represented by a vector followed by its opposite?

LESSON 1.6 IDENTIFYING VECTORS

In a previous lesson, you used **vectors** to add and subtract integers. A vector is modeled by an arrow that shows both magnitude and direction. The length of the arrow indicates the magnitude of the vector. The arrow indicates the direction of the vector. For example, the ship's velocity vector has magnitude 3 and direction to the right. This vector indicates that the ship's velocity is 3 miles per hour to the east.

3 miles/hr, east

Ongoing Assessment

Use a number line and draw a vector to show a move from 1 to 4. What is the length of the vector? What is the direction of the vector?

The arrow used to show the speed and direction of the ship could represent any vector with a magnitude of 3 and the same direction. For example, it could represent a force of three pounds pulling to the right. The vector could also represent a current flowing at three miles per hour downstream.

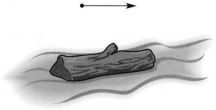

Force on Rope 3 lb, 45°N of E

River Current 3 mph, east

The way the arrow points tells you the direction of the vector. The length of the arrow tells you the magnitude or size. To determine magnitude, you must have a scale such as a number line or a number written next to the arrow.

Vectors can show the position of an object after two or more different moves.

Suppose a ship begins at point *A* and sails 10 miles north to point *B*. It then turns and sails 10 miles east to point *C*. Vectors *AB* and *BC* describe the two parts of the ship's journey. Vector *AB* has a magnitude of 10 miles and a direction of north. Vector *BC* also has a magnitude of 10 miles, but the direction is east. Why are vectors *AB* and *BC* *not* equal?

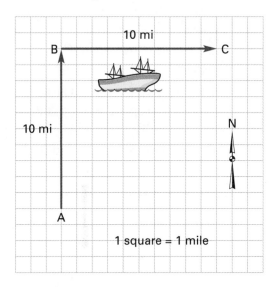

Vectors can model the addition and subtraction of several integers. Remember that subtracting an integer is the same as adding its opposite. The **opposite of a vector** is a vector that is the same size, but drawn in the opposite direction.

ACTIVITY   **Opposite Vectors**

1 Construct a number line.

2 Start at the origin and draw a vector to the right 4 units in magnitude.

3 Start at the origin and draw a vector to the left 4 units in magnitude.

4 Explain why these two vectors are called opposites.

Critical Thinking   Draw a vector representing a ship's movement 3 miles to the east. Then draw a vector to represent the opposite of the ship's vector. What is the result of a ship's journey represented by a vector followed by its opposite?

EXAMPLE Adding and Subtracting Integers

Use vectors to find $6 + (-4) - (-3)$.

SOLUTION

Start at 0 and draw a 6 vector.

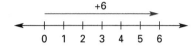

Start at 6 and draw a -4 vector.

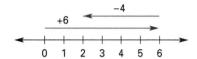

Start at 2 and draw the vector that is
the opposite of a -3 vector.

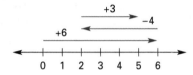

The total is 5.

Ongoing Assessment

Use vectors to find $-4 - 3 + (-5)$. Then explain how to find the
total without using vectors.

WORKPLACE COMMUNICATION

What is the "obvious" error in the
display of velocity vectors during
the test?

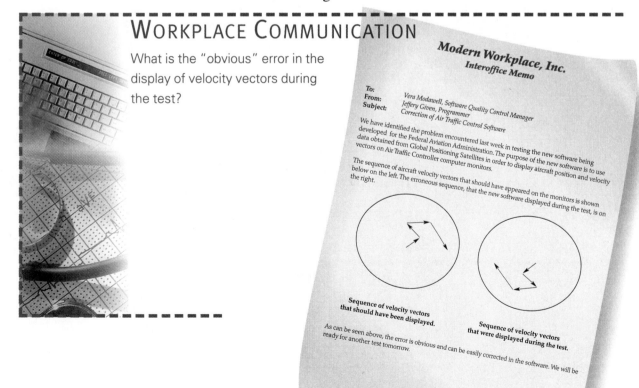

Modern Workplace, Inc.
Interoffice Memo

To: Vera Modawell, Software Quality Control Manager
From: Jeffery Given, Programmer
Subject: Correction of Air Traffic Control Software

We have identified the problem encountered last week in testing the new software being
developed for the Federal Aviation Administration. The purpose of the new software is to use
data obtained from Global Positioning Satellites in order to display aircraft position and velocity
vectors on Air Traffic Controller computer monitors.

The sequence of aircraft velocity vectors that should have appeared on the monitors is shown
below on the left. The erroneous sequence, that the new software displayed during the test, is on
the right.

Sequence of velocity vectors
that should have been displayed.

Sequence of velocity vectors
that were displayed during the test.

As can be seen above, the error is obvious and can be easily corrected in the software. We will be
ready for another test tomorrow.

LESSON ASSESSMENT

1 What two quantities are displayed by all vectors?

2 Draw a diagram showing a sailboat's movement if it sails 6 miles west and then turns south and sails 8 miles.

3 Draw the opposite of the vector that represents a journey 6 miles to the north.

4 Show how to use vectors to find $-7 + 4 - (-3)$. What is the total?

Practice and Problem Solving

Find the total using vectors.

5. $6 + (-5) + (-3)$

6. $-3 + 5 - (-2)$

7. $-6 - (-8) + 9$

8. $4 + 3 - (6)$

9. $-2 - (-6) - (-3)$

10. $0 - (-5) - 2$

Find the total without using vectors.

11. $-15 + 36 - 23$

12. $32 - (-35) + 18$

13. $-37 - (-12) - 28$

14. $43 - (-27) + (56)$

15. $51 + (-48) - (-52)$

16. $-23 - (-62) - (-36)$

17. $39 - (85) - (39)$

18. $-68 + (52) - (-84)$

19. $-16 - 37 + 15 - 28$

20. $48 - (-36) + (-72) - 68$

21. Mark opened a savings account for $75. He withdrew $30 on Wednesday and $20 on Friday. On the following Monday, Mark deposited $15. Represent each transaction with an integer. How much does Mark have in his savings account after the final deposit?

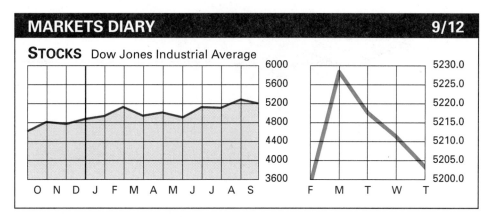

The chart shows the change in the stock market during four days' time. If the market started at 5220.25, give the value of the market at the close of each day.

22. Mon.; +8.5   **23.** Tue.; −11.125

24. Wed.; −6.375   **25.** Thu.; −7.75

26. Compare the changes in the stock market by drawing a number line and placing the changes for each day in the correct order. Which day has the biggest decline?

Combining Vectors

You can use vectors to help solve many problems.

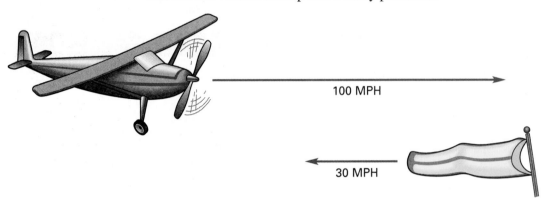

An airplane is flying with an airspeed of 100 miles per hour. The airplane is also flying directly into a headwind of 30 miles per hour. The plane does not actually travel 100 mph over the ground. To find the speed of the plane over the ground, the velocity vectors shown in the diagram are *combined*.

When you combine vectors, you need to think about the **absolute value** of each vector. The absolute value of a vector is its length or magnitude. For example, the length of the lower arrow represents the magnitude of the wind speed.

Absolute Value of a Vector
The absolute value or magnitude of a vector is the measure of its length.

Ongoing Assessment

What is the magnitude of the vector representing

a. the airplane speed?   **b.** the wind speed?

Now you can combine the velocity vectors for the airplane and the wind to find the velocity vector of the airplane over the ground.

1 On a piece of graph paper, make a scale drawing for the airplane and the wind velocity vectors. Let one square represent 10 miles per hour. Label your beginning point—the tail of the vector—as point A.

2 Draw a line segment 10 units long heading directly east from point A. Label the end point—the head of this vector—as point B. This vector represents the plane's velocity.

3 From point B—and a little below vector AB so that the lines do not overlap—draw another line segment 3 squares long heading due west. Label the head of this vector point C. This second vector represents the wind speed.

4 The result of adding these two vectors is the vector from point A, the beginning point, to point C, the ending point. How long is the vector from A to C? What is its direction? Compare your picture to the one below.

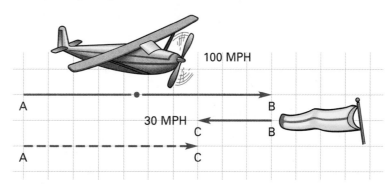

5 The direction of the vector from A to C tells you the direction of the plane. The length of the vector tells you the ground speed; that is, how fast the plane is actually passing over the ground. What is the airplane's velocity with respect to the ground?

Vectors at Right Angles

The velocity vectors in the airplane example point in opposite directions. Sometimes vectors are directed at right angles to each other. The drawing on the next page is similar to the one drawn for the ship example in Lesson 1.6. In this drawing, the vector AC is included with vectors AB and BC.

Use this drawing to complete Activity 2.

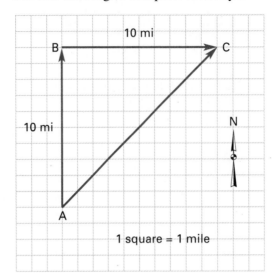

1 square = 1 mile

1 What is the scale for vectors *AB* and *BC*?

2 What is the direction of vector *AB*? What is its magnitude (in miles)?

3 What is the direction of vector *BC*? What is its magnitude (in miles)?

4 What is the direction of vector *AC*? Measure to find its approximate magnitude (in miles).

5 What does vector *AC* represent?

Combining Two Vectors

Justine and Jorge use ropes to pull a heavy cart. Jorge pulls to the left with a force of 50 pounds. Justine pulls at a 45-degree angle with a force of 30 pounds.

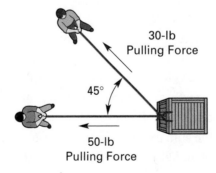

30-lb
Pulling Force

45°

50-lb
Pulling Force

You can draw a picture of what happens by drawing vectors of magnitude 50 pounds and 30 pounds in the directions of the two forces (along each rope). These two vectors are combined to find the resultant force on the cart.

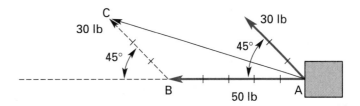

The method for combining the vectors is as follows:

First redraw the 30-pound vector with its tail at the head of the 50-pound vector. Notice that the 30 pounds is still directed at 45 degrees. Then draw the **resultant vector** from point *A* to point *C*, which shows the combined effect of the two separate forces.

ACTIVITY 3   Combining Forces

Use graph paper to make a scale drawing of the separate vectors representing Justine and Jorge pulling forces. Let each square on the graph paper represent 10 pounds of force.

1 Label your starting point *A*. Draw a vector 5 units long going directly to the left of point *A*. Label the arrowhead of this vector point *B*.

2 From point *B*, draw another vector 3 units long upward at a 45-degree angle from your first vector. Label the head of this vector point *C*. The result of adding these two vectors is the resultant vector *AC*.

3 Use a protractor to find the angle between the resultant vector and the vector *AB*.

4 Complete this statement.

Combining these two forces is the same as using one force of approximately _?_ pounds pulling at about _?_ degrees north of west.

LESSON ASSESSMENT

1 Research the meaning of each of the following terms. Explain how a vector can be used to represent the quantity.

   a. acceleration   **b.** displacement   **c.** velocity

2 How are two vectors combined to find the resultant vector?

   Describe a situation where vectors are used to represent

3 two forces acting in the same direction.

4 two forces acting in opposite directions.

5 two forces acting at a 90-degree angle to each other.

Practice and Problem Solving

Use graph paper to sketch vector diagrams for each situation.

6. A canoe floats due east on a river at a speed of 10 miles per hour across the water. The river current flows east at a speed of 3 miles per hour.

7. A canoe floats due south on a river at a speed of 10 miles per hour across the water. The river current flows to the south at a speed of 3 miles per hour.

A jet flew from Dallas to Atlanta with an airspeed of 400 miles per hour. The wind has a speed of 100 miles per hour as shown below.

8. Combine the two velocity vectors and sketch the resultant vector over the ground.

9. Is the jet's ground speed greater than or less than its air speed? Why?

On the return flight from Atlanta to Dallas, the velocity vector of the wind is the same.

10. Combine the two velocity vectors and sketch the resultant velocity vector of the jet over the ground.

11. Is the jet's ground speed greater than or less than its air speed? Why?

Mixed Review

The EZ-Duzzit Company budgeted a total of $5580 for Project Number 2468, to be completed over a span of six months. The management projected that the money would be equally spent during each of the six months. During the first four months, the following total amounts were reported spent: March $570, April $1071, May $1247, and June $928.

12. What is the budgeted monthly amount for each of the six months of Project Number 2468?

13. Make a table of monthly spending. For each month, compare the actual spending to the budgeted amount, and compute the amount "over/under budget." Represent each amount using an integer.

14. What is the total amount "over/under budget" for the first four months of the project?

The best selling card at City Card Shop is the Top Flight Card. During one 8-day period, the owner recorded the following daily sales of the Top Flight Card: 30, 41, 44, 50, 24, 32, 40, 43

Examining the data, the owner guessed the average daily sale for the Top Flight Card might be 40.

15. Subtract each actual daily sale from the owner's guessed average. Write the differences using integers.

16. Find the average of the integers you found in Exercise 15.

17. Subtract the integer you found in Exercise 16 from the guessed average.

18. Find the average of the daily sales by adding the sales and dividing by 8. How does this average compare to the number you found in Exercise 17?

19. Use the guessing method to find the average for the following set of test scores. 91, 94, 79, 80, 79, 75

MATH LAB

Equipment Timer
Calculator

Problem Statement

You will compare your pulse rate to the average pulse rate of your class following three different states of physical activity.

Procedure

a Locate your pulse. The two pulses you can find most easily are the radial pulse and the carotid pulse. The radial pulse is located on your wrist near the base of your thumb. The carotid pulse is on the side of your throat beside your jaw.

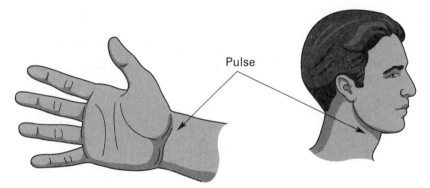

Pulse

Choose the location where your pulse is easiest for you to find.

b Find your pulse and count the number of beats in a 10-second period. Multiply this number by 6 to get the number of beats per minute. Write the number of beats per minute on a sheet of data paper as your "resting pulse rate."

c Run in place for one minute. Immediately count your pulse beats for a 10-second period. Multiply this number by 6 to get the beats per minute. Record the number of beats as your "exercising pulse rate."

d Rest for five minutes. Then count beats for a 10-second period. Multiply this number by 6 to get the beats per minute. Record this number as your "recovery pulse rate."

e On the chalkboard, write the headings for "resting pulse rate," "exercising pulse rate," and "recovery pulse rate." Record all the data for each class member on the chalkboard. Copy all the data to your record.

f Calculate the average pulse rate of the class for each state of activity. Use integers to report the difference between your pulse rate and the class average for each kind of rate.

g Use integers to report the increase from "resting pulse rate" to "exercising pulse rate" and the decrease from "exercising pulse rate" to "recovery pulse rate." Report these changes for your own pulse rates and for the average pulse rates for the class.

Activity 2: Drawing Displacement Vectors

Equipment   Directional compass
Calculator
Protractor
Tape measure
Two stakes

Problem Statement

A displacement vector is a distance traveled (magnitude) in a given direction. A distance of 50 feet at a compass heading of 45° (northeast) is an example of a displacement vector. You will move from a beginning point to an ending point by following directions given to you in terms of displacement vectors.

Procedure

a Select a location for one of the stakes and call it point A. Starting at point A, measure 20 feet along a compass heading of 55°. This locates point B. From point B, measure 15 feet along a compass heading of 145°. This locates point C. From point C, measure 10 feet along a compass heading of 265°. This locates point D. At point D, locate the other stake.

b Find the distance from point *A* to point *D*. Use your compass to determine the direction from *A* to *D*.

c Make a scale drawing using arrows as the displacement vectors in Step **a**. Draw a vector from point *A* to point *B*. From point *B*, draw a second vector to point *C*. From point *C*, draw a third vector to point *D*. Draw a final vector from *A* to *D* with the tail of the vector at *A* and the head at *D*. This vector is the resultant of the three displacement vectors.

d Based on the scale you used to make your drawing, determine the magnitude of the resultant vector.

e With a protractor, measure the heading (angle) of the resultant vector. Compare these results to the measured direction and distance from the beginning stake to the end stake. How do the two sets of measurements compare?

Activity 3: Combining Force Vectors

Equipment   Two spring scales with 500-gram capacities
Three golf tees
Calculator
Masking tape
Plumb line
Key ring
String (2 feet long)
500-gram hook weight
Protractor
Pegboard

Problem Statement

You can represent weight as a vector. The direction of this vector is usually toward the center of the Earth. You will "weigh" an object with two scales that are pulling at angles other than vertical. These two scales represent vectors. When the vectors are added together, they are equivalent to the weight of the object.

Procedure

a Tie the key ring 3 feet from the bob on the plumb line.

b Position three golf tees on the pegboard as shown in the drawing. Slip the key ring over the center tee. Tie a loop in

Suppose your travel alarm clock runs about one minute slow each day. On the back of the clock you see an adjustment lever in a slot.

(slow down) **−** ● ● ● ● ● ● ● ● ● ● **+** (speed up)

13 Which direction should you move the lever to adjust the speed?

14 If each dot above the slot represents a change of $\frac{1}{2}$ minute per day, by "how many dots" should you adjust your clock? Represent this adjustment as an integer.

A small airplane flies cross-country under various wind conditions. Through all the conditions, the plane maintains an air speed of 115 miles per hour and a compass heading due east. For each of the conditions, sketch the vectors representing the plane's air velocity (speed and direction), the wind velocity (speed and direction), and the ground velocity (speed and direction).

15 Wind from the west with a speed of 25 mph.

16 Wind from the east with a speed of 20 mph.

17 Wind from the south with a speed of 30 mph.

To run a cable across Murky Creek, you must determine the distance across the creek. Points A and D locate utility poles.

Start at point A. Walk due south for 21 meters. Then turn due east and walk 22.5 meters across the bridge. Finally, walk due north 9 meters to reach the desired point D.

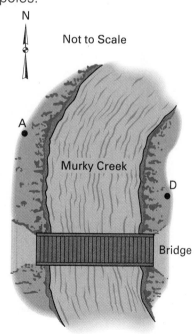

N

Not to Scale

Murky Creek

A

D

Bridge

18 Make a scale drawing of the distances and directions traveled. The final point should be labeled D.

19 The resultant vector joining point A to point D on your scale drawing is equivalent to the sum of the vectors that represent the distances traveled on each leg of the trip. Measure the magnitude of vector AD on your scale drawing. How many meters is it across Murky Creek from point A to point D?

each end of a 2-foot piece of string. Slip one loop over the hook at the end of one of the scales. Slip the other loop over the hook on the other scale. Hang the two scale rings on the two outer tees. Hook the 500-gram weight over the string at a point $\frac{1}{3}$ of the way from one end of the 2-foot string, then tape the hook neatly to the string with masking tape. Adjust the angle of the pegboard so that the plumb line crosses the 2-foot string at the position where the weight is hooked. The completed setup is shown in the drawing. Be sure that the plane of the pegboard is near-vertical and perpendicular to the floor. Do not let the scales "drag" against the pegboard.

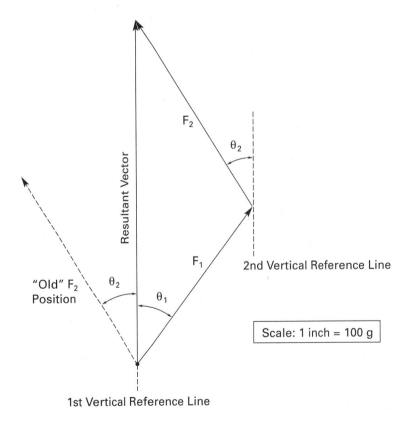

Pegboard

Golf Tees

Key Ring

500-g Scale

θ_2 θ_1

500-g Weight

Plumb Line

Resultant Vector

F_2

θ_2

F_1

2nd Vertical Reference Line

"Old" F_2 Position

θ_2

θ_1

Scale: 1 inch = 100 g

1st Vertical Reference Line

c Read the weight shown on each of the scales. On a sheet of data paper, record the weight from the scale on the right as F_1. Record the weight from the scale on the left as F_2.

d Measure the angle between the plumb line and the scale on the right (force F_1) and write this value on a sheet of data paper as θ_1. Measure the angle between the plumb line and the scale on the left (force F_2) and write the value on a sheet of paper as θ_2.

e Move the 500-gram weight to the halfway point of the 2-foot string and repeat Steps **b, c,** and **d**.

f Move the weight to a point $\frac{1}{4}$ of the way along the 2-foot string and repeat Steps **b, c,** and **d**. This gives you three different pairs of weight vectors. Each pair of weight vectors is equivalent to the suspended 500-gram weight.

g Draw the first vector pair. Begin by drawing a vertical reference line. Measure the angle θ_1. The drawing will help you see how to do this. Use a scale of 1 inch to 100 grams and draw the force F_1. Put an arrowhead on the end of the vector (away from the reference line) to show the direction of the vector. Now draw another vertical reference line through the arrowhead of F_1. Measure the angle θ_2 from the second vertical reference line and draw the vector for the force F_2 beginning at the head of F_1. Finally, draw a vector from the tail of F_1 to the head of F_2. This is the resultant vector. It represents the combined effects of F_1 and F_2 and should be equal, in scaled units, to the 500-gram weight. The resultant vector should be 5 inches long and drawn along the first vertical reference line.

h Draw vector diagrams for the other two vector pairs. Compare the three vector diagrams. For each resultant vector drawn, the length should be 5 inches, and the vector should point along the first vertical reference line. Do your three drawings show this to be true?

MATH APPLICATIONS

The applications that follow are like the ones you will encounter in many workplaces. Use the mathematics you have learned in this chapter to solve the problems. Wherever possible, use your calculator to solve the problems that require numerical answers.

The altitude of Owens Telescope Peak in southern California is 3367 meters above sea level. Just a few kilometers away, the floor of Death Valley is 86 meters below sea level.

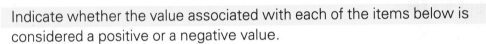

1 Let sea level be represented by 0. Express the altitude of each location as an integer.

2 What is the gain of altitude for a helicopter pilot who flies from the floor of Death Valley to the top of Owens Telescope Peak?

Indicate whether the value associated with each of the items below is considered a positive or a negative value.

3 A tax increase

4 Price reduction during a sale

5 The change in your checking account balance after you write a check

6 Temperature change when a cold front "passes through"

7 Stock market index rising 3 points

8 Difference in price paid for an item, after manufacturer's rebate

9 Change in pulse rate from a resting position to jogging

10 Tide at 2 feet above normal

11 Discounted airfares

12 A plunge in the wholesale price index

A business firm buys a second building across the street from its current location. Most of the office space is located on the second floor of each building. A raised walkway is proposed to reduce the walking time from one office area to the other. An overhead view of both the proposed raised walkway and the current route is shown in the drawing.

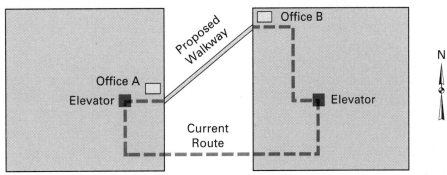

20 How many vectors are required to describe the path taken from office A to office B through hallways and doorways along the current route? (Remember the elevator rides between the second and first floors!)

21 What single vector can be used to represent the sum of these individual vectors?

22 Assume the distance from the beginning of one "dash" to the next is 10 feet. Make a scale drawing of the addition of the vectors representing the current route. Omit the elevator rides. About how long is the proposed walkway?

23 If the current route requires 10 minutes (assume the two elevator rides last 2 minutes combined), estimate the time for the trip across the walkway.

A hardware supply business keeps parts in small bins along one wall. You are looking for a particular part. A clerk who sees you looking in bin A says, " . . . no problem! From the bin you're now looking in, go down three rows and over five bins."

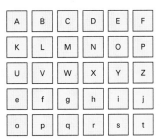

24 Make a quick sketch of the bins. On your sketch, draw the two vectors described by the clerk. What is the letter of the box that holds your part?

25 Write another combination of vectors to describe the location of the bin containing your part.

26 Draw the vector representing the sum of the two vectors in Exercise 24 and Exercise 25. How can you describe this resultant vector? Of the two methods to locate your part (the two vectors described by the clerk, or the one vector sum), which is the better way to describe the location of the bin you wanted?

AGRICULTURE & AGRIBUSINESS

An experimental process for increasing the growth rate of wheat is being tested and compared to the results predicted by many months of computer simulation. The predicted and actual values of growth rate increase for several trials of this process have been recorded in a table.

Test Results for Wheat Growth Rate Increases		
	Predicted Percent Increase	**Actual Percent Increase**
Trial A	1.103	1.007
Trial B	4.198	4.225
Trial C	14.611	14.803
Trial D	8.014	7.891

27 To evaluate the results of the test, you must compare the actual values to the predicted values. Copy the table. Add a column to show the difference "Actual − Predicted." Use a positive or negative number to show each difference.

28 You need to report the average difference for the test. Add the differences and divide by the number of trials.

29 Find the average using the absolute values of the differences. How does this approach differ from Exercise 28?

The average rainfall for one county is 0.8 inches per week during this time of year. The rainfall for the past five weeks is recorded in a table.

Weekly Rainfall	
Week 1	0.0"
Week 2	1.0"
Week 3	0.5"
Week 4	0.8"
Week 5	0.1"

30 Write the positive and negative difference of each week's rainfall from the average for this time of year. This difference is called a **deviation**.

31 What is the meaning of a positive deviation? A negative deviation?

32 What is the total deviation for this period of five weeks? Is the rainfall above normal or below normal?

When you apply insecticide with a sprayer, you must calibrate your applicator. If the application rate is higher or lower than the recommended amount, you must make an adjustment. A common method of testing the application rate is to hang small bags to collect the insecticide at a set distance and speed. The amounts from four trials at different application rates are recorded in a table. According to your calculations, each sample should have 9 ounces.

Calibration Samples	
Trial	**Sample amount**
1	7 oz
2	12 oz
3	10 oz
4	9 oz

33 For each sample, compute the difference between the actual amount and the target amount of 9 ounces. Which trial was farthest from the target application rate?

A newspaper reported the performance of various investments since the beginning of the year.

BUSINESS & MARKETING

Results of Various Investments This Year			
Investment	**Dec 31**	**Today**	**Percent change**
Japanese yen	121.00	130.95	+8.2%
Long-term T-bonds	2658.27	2816.47	+6.0%
Dow Industrials	1938.83	1983.26	+2.3%
Silver (ounce)	6.68	6.27	−4.6%
Gold (ounce)	486.20	445.30	−8.4%

34 Which investments have increased in value?

35 Which investments have decreased in value?

36 Which investment had the greatest increase?

37 Which investment had the greatest loss?

38 A monthly government report indicates a change from last month in the gross national product (GNP) of −1.8%. Does this change reflect a growth or a shrinkage of the GNP as compared to last month?

The net income/loss for Rocky Road Corporation is shown with a bar graph.

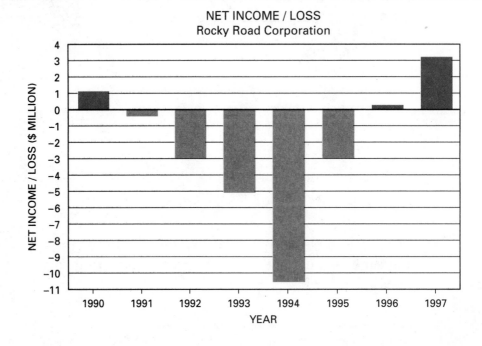

NET INCOME / LOSS
Rocky Road Corporation

39 Were most of the years shown in the graph losses or profits for Rocky Road Corporation?

40 Which year had the greatest income?

41 Which year had the greatest loss?

42 Do the recent years' net income/loss (the last 3–4 years) show a trend that is improving or worsening?

each end of a 2-foot piece of string. Slip one loop over the hook at the end of one of the scales. Slip the other loop over the hook on the other scale. Hang the two scale rings on the two outer tees. Hook the 500-gram weight over the string at a point $\frac{1}{3}$ of the way from one end of the 2-foot string, then tape the hook neatly to the string with masking tape. Adjust the angle of the pegboard so that the plumb line crosses the 2-foot string at the position where the weight is hooked. The completed setup is shown in the drawing. Be sure that the plane of the pegboard is near-vertical and perpendicular to the floor. Do not let the scales "drag" against the pegboard.

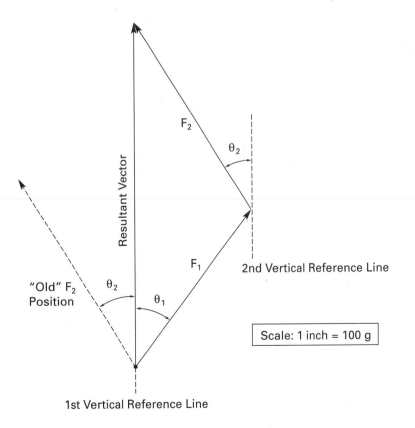

c Read the weight shown on each of the scales. On a sheet of data paper, record the weight from the scale on the right as F_1. Record the weight from the scale on the left as F_2.

d Measure the angle between the plumb line and the scale on the right (force F_1) and write this value on a sheet of data paper as θ_1. Measure the angle between the plumb line and the scale on the left (force F_2) and write the value on a sheet of paper as θ_2.

e Move the 500-gram weight to the halfway point of the 2-foot string and repeat Steps **b, c,** and **d.**

f Move the weight to a point $\frac{1}{4}$ of the way along the 2-foot string and repeat Steps **b, c,** and **d.** This gives you three different pairs of weight vectors. Each pair of weight vectors is equivalent to the suspended 500-gram weight.

g Draw the first vector pair. Begin by drawing a vertical reference line. Measure the angle θ_1. The drawing will help you see how to do this. Use a scale of 1 inch to 100 grams and draw the force F_1. Put an arrowhead on the end of the vector (away from the reference line) to show the direction of the vector. Now draw another vertical reference line through the arrowhead of F_1. Measure the angle θ_2 from the second vertical reference line and draw the vector for the force F_2 beginning at the head of F_1. Finally, draw a vector from the tail of F_1 to the head of F_2. This is the resultant vector. It represents the combined effects of F_1 and F_2 and should be equal, in scaled units, to the 500-gram weight. The resultant vector should be 5 inches long and drawn along the first vertical reference line.

h Draw vector diagrams for the other two vector pairs. Compare the three vector diagrams. For each resultant vector drawn, the length should be 5 inches, and the vector should point along the first vertical reference line. Do your three drawings show this to be true?

MATH APPLICATIONS

The applications that follow are like the ones you will encounter in many workplaces. Use the mathematics you have learned in this chapter to solve the problems. Wherever possible, use your calculator to solve the problems that require numerical answers.

The altitude of Owens Telescope Peak in southern California is 3367 meters above sea level. Just a few kilometers away, the floor of Death Valley is 86 meters below sea level.

1 Let sea level be represented by 0. Express the altitude of each location as an integer.

2 What is the gain of altitude for a helicopter pilot who flies from the floor of Death Valley to the top of Owens Telescope Peak?

Indicate whether the value associated with each of the items below is considered a positive or a negative value.

3 A tax increase

4 Price reduction during a sale

5 The change in your checking account balance after you write a check

6 Temperature change when a cold front "passes through"

7 Stock market index rising 3 points

8 Difference in price paid for an item, after manufacturer's rebate

9 Change in pulse rate from a resting position to jogging

10 Tide at 2 feet above normal

11 Discounted airfares

12 A plunge in the wholesale price index

Suppose your travel alarm clock runs about one minute slow each day. On the back of the clock you see an adjustment lever in a slot.

(slow down) **—** • • • • • • • • • **+** (speed up)

13 Which direction should you move the lever to adjust the speed?

14 If each dot above the slot represents a change of $\frac{1}{2}$ minute per day, by "how many dots" should you adjust your clock? Represent this adjustment as an integer.

A small airplane flies cross-country under various wind conditions. Through all the conditions, the plane maintains an air speed of 115 miles per hour and a compass heading due east. For each of the conditions, sketch the vectors representing the plane's air velocity (speed and direction), the wind velocity (speed and direction), and the ground velocity (speed and direction).

15 Wind from the west with a speed of 25 mph.

16 Wind from the east with a speed of 20 mph.

17 Wind from the south with a speed of 30 mph.

To run a cable across Murky Creek, you must determine the distance across the creek. Points A and D locate utility poles.

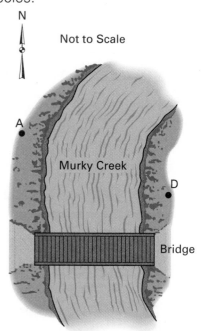

Start at point A. Walk due south for 21 meters. Then turn due east and walk 22.5 meters across the bridge. Finally, walk due north 9 meters to reach the desired point D.

18 Make a scale drawing of the distances and directions traveled. The final point should be labeled D.

19 The resultant vector joining point A to point D on your scale drawing is equivalent to the sum of the vectors that represent the distances traveled on each leg of the trip. Measure the magnitude of vector AD on your scale drawing. How many meters is it across Murky Creek from point A to point D?

A business firm buys a second building across the street from its current location. Most of the office space is located on the second floor of each building. A raised walkway is proposed to reduce the walking time from one office area to the other. An overhead view of both the proposed raised walkway and the current route is shown in the drawing.

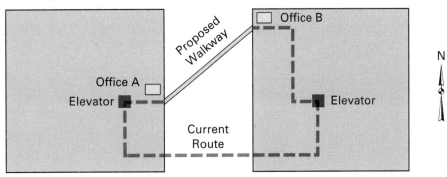

20 How many vectors are required to describe the path taken from office A to office B through hallways and doorways along the current route? (Remember the elevator rides between the second and first floors!)

21 What single vector can be used to represent the sum of these individual vectors?

22 Assume the distance from the beginning of one "dash" to the next is 10 feet. Make a scale drawing of the addition of the vectors representing the current route. Omit the elevator rides. About how long is the proposed walkway?

23 If the current route requires 10 minutes (assume the two elevator rides last 2 minutes combined), estimate the time for the trip across the walkway.

A hardware supply business keeps parts in small bins along one wall. You are looking for a particular part. A clerk who sees you looking in bin A says, " . . . no problem! From the bin you're now looking in, go down three rows and over five bins."

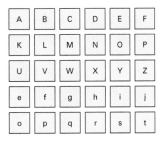

24 Make a quick sketch of the bins. On your sketch, draw the two vectors described by the clerk. What is the letter of the box that holds your part?

25 Write another combination of vectors to describe the location of the bin containing your part.

26 Draw the vector representing the sum of the two vectors in Exercise 24 and Exercise 25. How can you describe this resultant vector? Of the two methods to locate your part (the two vectors described by the clerk, or the one vector sum), which is the better way to describe the location of the bin you wanted?

AGRICULTURE &
AGRIBUSINESS

An experimental process for increasing the growth rate of wheat is being tested and compared to the results predicted by many months of computer simulation. The predicted and actual values of growth rate increase for several trials of this process have been recorded in a table.

Test Results for Wheat Growth Rate Increases		
	Predicted Percent Increase	Actual Percent Increase
Trial A	1.103	1.007
Trial B	4.198	4.225
Trial C	14.611	14.803
Trial D	8.014	7.891

27 To evaluate the results of the test, you must compare the actual values to the predicted values. Copy the table. Add a column to show the difference "Actual − Predicted." Use a positive or negative number to show each difference.

28 You need to report the average difference for the test. Add the differences and divide by the number of trials.

29 Find the average using the absolute values of the differences. How does this approach differ from Exercise 28?

The average rainfall for one county is 0.8 inches per week during this time of year. The rainfall for the past five weeks is recorded in a table.

Weekly Rainfall	
Week 1	0.0"
Week 2	1.0"
Week 3	0.5"
Week 4	0.8"
Week 5	0.1"

30 Write the positive and negative difference of each week's rainfall from the average for this time of year. This difference is called a **deviation**.

31 What is the meaning of a positive deviation? A negative deviation?

32 What is the total deviation for this period of five weeks? Is the rainfall above normal or below normal?

When you apply insecticide with a sprayer, you must calibrate your applicator. If the application rate is higher or lower than the recommended amount, you must make an adjustment. A common method of testing the application rate is to hang small bags to collect the insecticide at a set distance and speed. The amounts from four trials at different application rates are recorded in a table. According to your calculations, each sample should have 9 ounces.

Calibration Samples	
Trial	Sample amount
1	7 oz
2	12 oz
3	10 oz
4	9 oz

33 For each sample, compute the difference between the actual amount and the target amount of 9 ounces. Which trial was farthest from the target application rate?

A newspaper reported the performance of various investments since the beginning of the year.

BUSINESS & MARKETING

Results of Various Investments This Year			
Investment	Dec 31	Today	Percent change
Japanese yen	121.00	130.95	+8.2%
Long-term T-bonds	2658.27	2816.47	+6.0%
Dow Industrials	1938.83	1983.26	+2.3%
Silver (ounce)	6.68	6.27	−4.6%
Gold (ounce)	486.20	445.30	−8.4%

34 Which investments have increased in value?

35 Which investments have decreased in value?

36 Which investment had the greatest increase?

37 Which investment had the greatest loss?

38 A monthly government report indicates a change from last month in the gross national product (GNP) of −1.8%. Does this change reflect a growth or a shrinkage of the GNP as compared to last month?

The net income/loss for Rocky Road Corporation is shown with a bar graph.

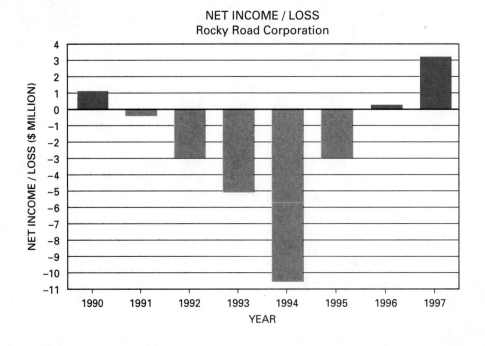

NET INCOME / LOSS
Rocky Road Corporation

39 Were most of the years shown in the graph losses or profits for Rocky Road Corporation?

40 Which year had the greatest income?

41 Which year had the greatest loss?

42 Do the recent years' net income/loss (the last 3–4 years) show a trend that is improving or worsening?

The total cost of merchandise sold in a clothing store is computed by adding or subtracting the following amounts.

Total cost of merchandise	=	+ cost of beginning inventory
		+ billed cost of net purchases
		+ inbound transportation cost
		− cost of ending inventory
		− cash discounts
		+ workroom/alteration costs

43 A clothing store showed the following figures for a six-month period: beginning inventory, $13,651; billed cost of net purchases, $87,055; transportation charges $4,879; cash discounts earned on purchases, $7,510; cost of ending inventory $12,402; workroom costs, $748. Compute the total cost of the merchandise sold.

The Smooth-Zipper Corporation publishes a balance sheet that compares this year's performance with last year's. Here is a portion of the balance sheet.

Smooth-Zipper Corp.
Balance Sheet
December 31

Assets	Last year	This year	Increase/Decrease
Cash	$12,500	$13,800	
Inventory	42,400	47,500	
Accounts receivable	24,700	22,000	
Land	$12,000	$12,500	
Buildings	35,000	37,200	
Equipment	8,600	8,000	
Total assets	$135,200	$141,000	

44 Complete the last column of the statement to report the increase or decrease for each asset.

45 What is the difference in value between the assets for this year and last year? Compare the sum of the third column of integers with the difference in total assets.

During the week, the warehouse for Comfy Footwear posted these adjustments to the inventory.

Inventory Adjustments for Week Ending 2/29	
Date	Adjustment
2/24	+16
2/25	−28
2/25	−2
2/26	+1
2/26	−3
2/27	+12
2/28	+8
2/29	−14

46 What is the total of the adjustments made for this week?

47 The inventory at the beginning of the week is 18,384 units. By the end of the day on 2/29, 720 units have been shipped from the warehouse. Allowing for the adjustments, what is the inventory at the end of this week?

An office manager is preparing a report that compares performance to schedule for the past quarter. A list of projects and their completion dates are recorded in a table.

Project Schedule Summary		
Project ID	Scheduled completion date	Actual completion date
1010	3/30	3/29
4242	3/10	3/10
5142	2/17	2/15
6231	3/25	3/28
6323	2/15	2/12
7652	1/29	1/28
7774	2/8	2/10

48 Find the difference (number of days) between the scheduled completion date and the actual completion date. Early project completion is represented by a negative integer. Record each difference as an integer.

49 How many days were lost or gained by all the projects for the quarter?

You have purchased 144 pairs of earrings to sell in your store. You are going to sell each pair at a retail price of $4.98. During the first 30 days, you sold 60 pairs. Your total expenses for selling the 60 pairs totaled $280.87. You sold the remainder during the next 30 days, when the expenses for these pairs were $339.24.

50 Compute the total retail value of the earrings sold during the first 30 days and during the second 30 days.

51 Subtract the expenses from the total retail profit for each of the two periods to find the net profit (or loss).

52 Find the total net profit or loss by adding the net profit from each 30-day period.

A market research study for River City reported the median income for various age groups. You decide to compare this report with a similar report done five years ago. During the past five years, wages should have increased about 22% to keep up with inflation. The table shows the results of the two studies.

Comparison of Market Research Studies for River City Median Annual Incomes		
Age group	**Previous study**	**Current study**
12–18	$6,000	$7,000
19–25	$17,500	$21,500
26–45	$47,500	$58,500
46–65	$49,000	$59,000
66 and above	$33,500	$41,500

Copy the table. Add a third column to your table with the heading, "Projected Income."

53 For each age group, calculate the projected income necessary to keep pace with 22% inflation over the five years. Record each value.

54 Compare the projected incomes with the results of the current study. Add a fourth column to your table showing the difference between the current and the projected incomes. Record each difference as an integer.

55 Which sentence describes your use of a negative integer?

a. Income has not kept pace with inflation.
b. Income has exceeded the value projected by the inflation rate?

56 Which age group(s) in River City is within $500 of its projected income?

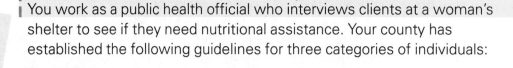

You work as a public health official who interviews clients at a woman's shelter to see if they need nutritional assistance. Your county has established the following guidelines for three categories of individuals:

Required Nutrition (servings per day)

Food group	Children	Pregnant women	Other women
Fruits and vegetables	4	4	4
Breads and grains	4	4	4
Meats and other protein sources	2	2	2
Dairy products	3	4	2

Each client's eating habits are recorded. If a total of three or more deficiencies is discovered, the client is eligible for assistance. For example, one missed serving in a food group is one deficiency, two missed servings is two deficiencies, and so on. Excess servings in a food group have no affect on eligibility.

You are interviewing a pregnant woman who has one small child. You have recorded the eating habits of the mother and child.

Servings per Day

Food group	Child	Pregnant mother
Fruits and vegetables	2	3
Breads and grains	5	3
Meats and other protein sources	1	2
Dairy products	3	2

57 Compute the difference between the actual and recommended number of servings per day for both the mother and the child. Record the differences as integers.

58 Record the total number of deficiencies for both the mother and the child.

59 Use the guidelines to determine if either the mother or the child is eligible for assistance.

You have purchased 144 pairs of earrings to sell in your store. You are going to sell each pair at a retail price of $4.98. During the first 30 days, you sold 60 pairs. Your total expenses for selling the 60 pairs totaled $280.87. You sold the remainder during the next 30 days, when the expenses for these pairs were $339.24.

50 Compute the total retail value of the earrings sold during the first 30 days and during the second 30 days.

51 Subtract the expenses from the total retail profit for each of the two periods to find the net profit (or loss).

52 Find the total net profit or loss by adding the net profit from each 30-day period.

A market research study for River City reported the median income for various age groups. You decide to compare this report with a similar report done five years ago. During the past five years, wages should have increased about 22% to keep up with inflation. The table shows the results of the two studies.

Comparison of Market Research Studies for River City Median Annual Incomes		
Age group	**Previous study**	**Current study**
12–18	$6,000	$7,000
19–25	$17,500	$21,500
26–45	$47,500	$58,500
46–65	$49,000	$59,000
66 and above	$33,500	$41,500

Copy the table. Add a third column to your table with the heading, "Projected Income."

53 For each age group, calculate the projected income necessary to keep pace with 22% inflation over the five years. Record each value.

54 Compare the projected incomes with the results of the current study. Add a fourth column to your table showing the difference between the current and the projected incomes. Record each difference as an integer.

55 Which sentence describes your use of a negative integer?

a. Income has not kept pace with inflation.
b. Income has exceeded the value projected by the inflation rate?

56 Which age group(s) in River City is within $500 of its projected income?

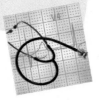

HEALTH OCCUPATIONS

You work as a public health official who interviews clients at a woman's shelter to see if they need nutritional assistance. Your county has established the following guidelines for three categories of individuals:

Required Nutrition (servings per day)			
Food group	Children	Pregnant women	Other women
Fruits and vegetables	4	4	4
Breads and grains	4	4	4
Meats and other protein sources	2	2	2
Dairy products	3	4	2

Each client's eating habits are recorded. If a total of three or more deficiencies is discovered, the client is eligible for assistance. For example, one missed serving in a food group is one deficiency, two missed servings is two deficiencies, and so on. Excess servings in a food group have no affect on eligibility.

You are interviewing a pregnant woman who has one small child. You have recorded the eating habits of the mother and child.

Servings per Day		
Food group	Child	Pregnant mother
Fruits and vegetables	2	3
Breads and grains	5	3
Meats and other protein sources	1	2
Dairy products	3	2

57 Compute the difference between the actual and recommended number of servings per day for both the mother and the child. Record the differences as integers.

58 Record the total number of deficiencies for both the mother and the child.

59 Use the guidelines to determine if either the mother or the child is eligible for assistance.

When a patient is recovering from an illness, it is important to monitor the patient's fluid intake and output. A fluid imbalance must be remedied as soon as possible. A record of a patient's fluids for several 8-hour periods during a hospital stay are recorded in a table.

Intake-Output Record for J. Smith			
Date	Shift	Intake	Output*
3/10	A	1200 cc	900 cc
3/10	B	1000 cc	800 cc
3/10	C	1050 cc	950 cc
3/11	A	1300 cc	1250 cc
3/11	B	1500 cc	1600 cc
3/11	C	1200 cc	1300 cc
3/12	A	1100 cc	1250 cc

*Includes estimated perspiration, etc.

60 A fluid balance is indicated by a difference between the intake and output of nearly zero. Make a table of the differences between the intake and output for each shift shown in the table. Use integers to record your differences.

61 Which of your differences indicates an imbalance due to insufficient output of fluids?

62 Are the absolute values of these differences useful in spotting a condition of fluid imbalance? What information have you lost if you show only the absolute values of the differences on your chart?

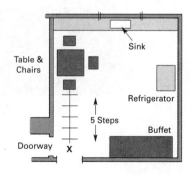

To help a person who has recently lost eyesight orient himself in a new room, you may describe the location of some of the important objects in that room. One method is to use clock positions. 12 o'clock is straight ahead, 3 o'clock is to the right, etc. Consider the room shown in the illustration. Notice the location of the labeled objects. The vision-impaired person is standing in the doorway at the point labeled X, facing the table and chairs.

63 Trace the basic room layout. Notice the scale provided, in terms of "steps." Draw the vectors that show the steps and direction from X needed to reach each of the labeled objects or locations:

a. Nearest chair
b. Sink
c. Refrigerator
d. Nearest edge of buffet
e. Other doorway (assume person enters from doorway at bottom of drawing)

64 Write the words that describe each of these vectors. Use the clock position to communicate the direction of each vector.

Eyeglass prescriptions define the amount of curvature required in a lens to correct a patient's vision. The amount of curvature or sphere power is expressed in units of diopters (D). To correct a near-sighted condition requires a negative value for sphere power, while to correct a far-sighted condition requires a positive value. A value close to zero indicates very little correction. Examine the values of sphere power listed below.

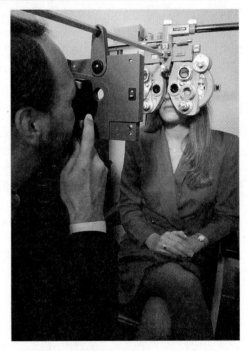

Values of Sphere Power Prescribed	
Patient	Sphere Power (D)
A	−2.00
B	+0.50
C	+1.00
D	−1.25
E	−3.75
F	+2.00

65 Draw a number line. For each patient, label the position on your number line representing the sphere power prescribed. Label the range of corrections for the near-sighted patients, and the range of correction for far-sighted patients.

66 Which patient requires the most correction for near-sightedness? For far-sightedness?

Some tax forms from the Internal Revenue Service have specific instructions for filling them out. One form has the following instruction for line 43.

Subtract line 42 from line 39.
Enter the result (but not less than zero).

For each set of values for line 42 and line 39 shown below, determine the correct value for line 43.

67 Line 39: $12,600
Line 42: $4267

68 Line 39: $2328
Line 42: $2230

69 Line 39: $3125
Line 42: $3419

You are balancing your checkbook with your calculator. Your starting balance is $264.88. You have written checks for $38.76, $144.69, $51.25, and $30.03. Your last bank statement showed that $5.16 interest was credited to your account and $2.50 was deducted for a service charge.

70 Make a list of the transactions posted to your account. Indicate subtractions from your account balance as negative numbers. Indicate additions to your account as positive numbers.

71 What is the final balance of your account after these transactions? What does the sign of the number tell you?

You are a flight attendant who travels worldwide with an international airline. As you fly, you experience many changes in local time zones. You have discovered that the local time increases when you travel east (+) and decreases when you travel west (−). The table shows some of the cities your airline travels to and the difference in local time compared to the time in London.

Location	Hours Changed
Sydney, Australia	+10
Tokyo, Japan	+ 9
Hong Kong, China	+ 8
Tehran, Iran	+ 4
Moscow, Russia	+ 3
Helsinki, Finland	+ 2
Rome, Italy	+ 1
London, England	0
Rio de Janeiro, Brazil	− 3
Halifax, Nova Scotia	− 4
New York, New York, USA	− 5
Chicago, Illinois, USA	− 6
Denver, Colorado, USA	− 7
Los Angeles, California, USA	− 8
Honolulu, Hawaii, USA	−10

72 Suppose you leave London and travel to Helsinki. How should you adjust your watch when you arrive in Helsinki?

73 The next day you fly from Helsinki to Moscow and then on to Rome. How should you adjust your watch when you arrive in Rome?

74 Soon after this, you leave Rome, fly west to New York and then on to Denver. How should you adjust your watch when you arrive in Denver?

You have received a monthly credit card statement with a summary of your transactions.

+	Previous Balance	$294.39
−	Payments	200.00
−	Credits	17.81
+	Purchases and cash advances	128.33
+	Debit adjustments	0.00
+	Finance charge	1.87
	* New Balance	$206.78

* An amount followed by a minus sign is a credit

75 How much did you pay during the last month to reduce your credit card debt? Why is it listed as a negative number?

76 What is the meaning of a positive value in the summary?

77 What circumstances would create a condition referred to by the footnote ("* An amount followed . . . ")?

INDUSTRIAL TECHNOLOGY

An automobile's ammeter shows whether the battery is charging (+) or discharging (−), and the current in the circuit. Examine the ammeter displays.

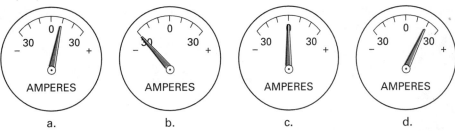

a.   b.   c.   d.

78 Write the reading from each display. Use the sign to indicate a charging or a discharging condition.

79 If a fuse fails, it will do so at the maximum current. Write the absolute value of the current readings from Exercise 78. Which ammeter reading shows the current with the greatest absolute value?

A furniture manufacturing line accepts hardwood lumber that has a specification (spec) weight of 20.7 pounds. When the lumber pieces enter the plant, they proceed down a conveyor belt and pass over a weigh scale set to the spec weight. You view the window of the weigh scale as each piece passes and record the deviations from "spec."

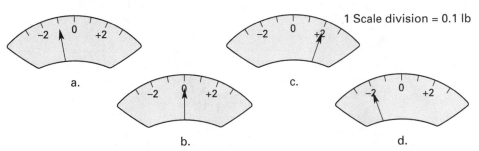

1 Scale division = 0.1 lb

a.   c.

b.   d.

80 Examine the weigh scale windows. Record each deviation from spec.

81 Find the average deviation from spec weight for your observations.

82 Use the average deviation to determine the average weight of the four pieces you observed.

Suppose a tank is filled at the rate of 40 gallons of water per minute. Meanwhile, a valve is opened to drain 32 gallons per minute into a pipeline.

83 Express each rate as an integer.

84 Use these integers to compute the number of gallons of water flowing into the tank in 10 minutes and the number of gallons drained from the tank in 10 minutes.

85 If the tank starts out empty, what is the total amount of water in the tank after 10 minutes?

The ignition timing of an automobile is referenced to "top dead center," more commonly known as TDC. The specification for a certain car calls for a timing of 10° before TDC. An inspection of a customer's poorly running car reveals the ignition timing is actually firing at 2° *after* TDC.

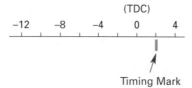

(TDC)

−12   −8   −4   0   4

Timing Mark

86 Express the specification timing and the actual timing as integers.

87 How much does the actual timing differ from the specification timing?

You and a coworker carry a tool chest that weighs 86.6 pounds. Using the handles on each end of the chest, the two of you carry the chest as shown in the drawing.

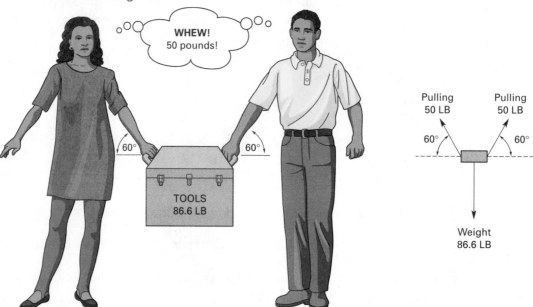

WHEW!
50 pounds!

60°   60°

TOOLS
86.6 LB

Pulling
50 LB   Pulling
                50 LB

60°   60°

Weight
86.6 LB

88 Suppose that each of you pulls with a force of 50 pounds at an angle of 60°, as shown in the drawing. What is the resultant (vector sum) of these two pulling forces? What is the effect of the resultant force on the tool chest?

89 What would the resultant vector be if you and your coworker each pulled with a force of 50 pounds at an angle of 45°? What is the effect of this resultant force on the tool chest?

90 If each of you lift straight up—at 90°—what is your lifting force on the tool chest? Show your answer with a vector diagram.

Examine the simplified side view of a metal framework requiring an interior weld. Note the initial position of the welding tip. You have been assigned the task of writing a program for the robotic welder. This progam will move the welding tip from the location shown to the weld location.

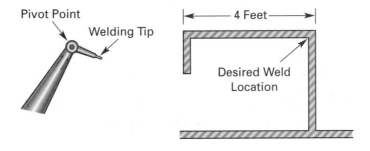

91 Sketch the metal framework and initial position of the welding tip. Draw a set of vectors your program could contain that would move the welding tip to the desired location without striking the frame. (Consider only the positioning of the tip. For this exercise, do not worry about the rest of the robot arm.)

92 How many different answers are there to this problem?

93 Draw the resultant vector that is the sum of your vector instructions. How many different answers are there to this question?

CHAPTER 1 ASSESSMENT

Skills

Find the absolute value.

1. $|-8|$ **2.** $|25|$ **3.** $|-16|$

Which integer is larger?

4. -6 or -3 **5.** -8 or -12 **6.** 0 or -3

Add, subtract, multiply, or divide.

7. 9×-6 **8.** $-23 - 51$ **9.** $-9 + (-5)$

10. $18 - (-14)$ **11.** $-36 \div -9$ **12.** $56 \div -7$

13. $-28 + 35$ **14.** $24 - (-17)$ **15.** -13×-6

16. $-21 \div 3$ **17.** $-16 - (-16)$ **18.** $-84 \div -4$

Applications

19. The stock market lost 35 points over 5 days. Write the average daily loss as an integer.

20. The record high temperature for a city is 107°F. The record low is −25°F. Find the difference between the high and low record temperatures.

The Bay Country Store keeps a record of daily sales. Use the table to complete the Exercises.

Day	Daily Sales Previous week's sales	This week's sales
Monday	$848	$812
Tuesday	$915	$975
Wednesday	$639	$612
Thursday	$986	$846
Friday	$743	$850

21. Write each daily sales difference using an integer.

22. What is the average difference in the daily sales?

Draw a vector diagram on graph paper for problems 23–26. Find the magnitude and direction of each resultant vector in problems 24–26.

23. One person pulls on a cart with a force of 65 pounds due east. Another person pulls with a force of 75 pounds in a direction 35° south of east.

24. An airplane flying east with an air speed of 435 miles per hour.

25. An airplane flying east with an air speed of 415 miles per hour with a *tail wind* of 15 miles per hour.

26. An airplane flying east with an air speed of 425 miles per hour into a *head wind* of 15 miles per hour.

Math Lab

27. Use integers to report the change in pulse rate from exercising at 155 beats per minute (bpm) to complete recovery at 68 bpm.

28. Use graph paper to find the resultant displacement vector when the following displacement vectors are added in the order shown.
> 30 feet north
> 50 feet west
> 15 feet south

29. Two spring scales are attached to a weight as shown in the diagram.

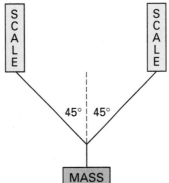

Each scale reads 353.5 grams. What is the weight of the mass?

CHAPTER 2

WHY SHOULD I LEARN THIS?

Scientific notation is a language used in today's high tech workplace. This chapter explores how these mathematical concepts are used in professions ranging from aeronautics to health care. Find out how scientific notation can be part of your future career.

SCIENTIFIC NOTATION

OBJECTIVES

1. Write large and small numbers in powers-of-ten notation.
2. Read and write numbers in scientific notation.
3. Use a scientific calculator to compute numbers displayed in scientific notation.
4. Use numbers written in scientific notation to solve problems.

Many people must use *very large* and *very small* numbers every day on the job. People who work in the space industry, in national defense, or at the World Bank often use large numbers. For example, astronomers look through telescopes to study stars and planets that are great distances from the Earth. Scientists and technicians working in a biology lab or a center for disease control use very small numbers when working with viruses and human cells.

People working in science related jobs use very large and very small number facts. For example,

An ounce of gold contains approximately
86,700,000,000,000,000,000,000 atoms.

One atom of gold has a mass of
0.000000000000000000000327 grams.

The average volume of an atom of gold is
0.000000000000000000000001695 cubic centimeters.

The mass of the earth is about
6,000,000,000,000,000,000,000 metric tons.

The diameter of a single human red blood cell is about
0.000007 meters.

In this chapter you will use a shortcut way to write very large and very small numbers. This shortcut is called *scientific notation*. As you view the video, watch for the use of very large and very small numbers. Notice how people use them at work.

LESSON 2.1 POWERS OF TEN

Exponents and Powers

You already know how to abbreviate the expression

$$2 + 2 + 2 + 2 + 2 = 10$$

by writing it as $5 \times 2 = 10$. There is also a way to abbreviate repeated multiplication. In the sentence

$$2 \times 2 \times 2 \times 2 \times 2 = 32$$

2 is repeated as a factor 5 times, and the product is 32. You can abbreviate this sentence by writing the number 5 up and to the right of the 2.

$$2^5 = 32$$

In this form, 2 is called the base, and 5 is called the exponent. This sentence is read "2 to the 5th power is 32."

You can raise any integer to a positive power.

Positive Exponent

A positive exponent tells you how many times to use the base as a factor in a repeated multiplication.

For example,

$$(-4)^3 = (-4) \times (-4) \times (-4) = -64$$

When you raise a negative number to a power, use parentheses to show the integer is used as a factor. Thus $(-4)^2 = 16$, but $-4^2 = -16$.

Some powers have special names. The third power is the cube of a number and the second power is the square of a number. The first power of a number is the number itself.

Ongoing Assessment

What is the 4th power of 2? What is -3 squared? What is 5 cubed?

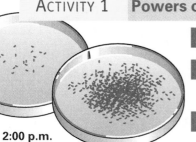

Critical Thinking   Suppose you raise a negative integer to a positive power. How can you tell if the resulting value is positive or negative?

There are usually several ways to raise a number to a power on a calculator. On most scientific calculators, there is an x-square key, x^2. There is also a power key, y^x, that raises any number to any power. Experiment with your calculator. Use the power key to check the answers to the assessment questions.

Most programmable calculators use the caret key, ^, to find powers. When you enter -2^4, the calculator displays 16.

Activity 1   Powers of 2

1   Make a table of the first 5 positive powers of 2.

2   Explain why the positive powers of two are sometimes referred to as "doubling powers."

3   Suppose the number of bacteria in a culture doubles each minute. If there are 2 bacteria in a culture at 2 PM, how many will be in the culture at 2:15 PM?

2:00 p.m.

2:15 p.m.

Positive Powers of Ten

The numbers 1000, 10,000, or 100,000 are all powers of ten. When you write powers of ten, you write the base 10 with an exponent. For example, the second power of 10, or 10 squared, is written 10^2.

Other examples

1000	=	10^3	base 10, exponent 3
10,000	=	10^4	base 10, exponent 4
100,000	=	10^5	base 10, exponent 5

In each case, the positive exponent tells you how many times to write down the factor 10 and multiply to get the number. You can see a pattern for writing powers of ten by examining the following table

$$10^1 = 10.$$
$$10^2 = 10 \times 10 = 100.$$
$$10^3 = 10 \times 10 \times 10 = 1000.$$
$$10^4 = 10 \times 10 \times 10 \times 10 = 10,000.$$

The decimal point moves one place to the right each time you multiply by another 10.

> **Multiplying by 10**
> Multiplying by 10 is the same as moving the decimal point one place to the right.

ACTIVITY 2 Powers of 10

Use the powers of ten pattern to complete each statement.

1 To write 10^4 as an integer, place __?__ zeros after the 1.

2 To write 10^3 as an integer, place __?__ zeros after the 1.

3 To write 10^2 as an integer, place __?__ zeros after the 1.

4 What is the relationship between the exponent and the number of zeros after the 1?

5 Extend and complete the table for 10^5 and 10^6.

Now do the problem "backward." First write down the integer and then find the power of ten it equals. The first line in the following table is done to get you started. Finish the other two lines.

$$\overset{\text{6 zeros}}{1{,}000{,}000} = \overset{\text{6 tens}}{10 \times 10 \times 10 \times 10 \times 10 \times 10} = \overset{\text{exponent is 6}}{10^6}$$

6 $10{,}000 = $ _____?_____ $= $ __?__

7 $100 = $ _____?_____ $= $ __?__

8 Explain the relationship between the number of zeros after the 1 and the exponent of 10.

LESSON ASSESSMENT

Think and Discuss

1 What is the meaning of an exponent?

2 How do you find the third power of a positive integer?

3 How do you know how many zeros follow the one when you find any power of ten?

4 How do you write ten million as a power of ten?

Practice and Problem Solving

Write each expression using an exponent.

5. 3×3

6. $2 \times 2 \times 2 \times 2 \times 2$

7. -5×-5

8. $10 \times 10 \times 10$

9. $-3 \times -3 \times -3 \times -3 \times -3$

10. $1 \times 1 \times 1$

Write each expression as an integer.

11. 3^3

12. 4^2

13. $(-1)^3$

14. 10^6

15. 2^7

16. $(-2)^3$

17. An electrical generation plant produces over 10^6 watts of power. Write this power level as an integer.

18. The distance between the Earth and the sun is about 10^8 kilometers. Write this distance as an integer.

19. Write 5×10^4 as an integer.

20. Write 7×10^6 as an integer.

Which is greater? Explain why.

21. 3^4 or 4^3

22. $(-2)^3$ or $(-1)^2$

23. 5^4 or 4^5

Mixed Review

Suppose you are keeping the weight loss statistics in pounds for a fitness club. The table shows the weekly gains and losses for five of your customers.

Person	Week 1	Week 2	Week 3	Week 4
Jose	−2	−1	−2	+1
Tim	−3	−2	−4	+1
Cathy	−4	−1	−2	−5
Maria	−2	+2	−2	+2
Pat	0	+2	0	+2

24. What is the total weight gain or loss for each customer?

+2   −3

25. Draw a number line and place each person in order according to weight gain or loss.

26. What is the average weight gain or loss per week for each customer?

27. What is the average weight gain or loss per week for the group?

28. Compare the average gain or loss per week for each customer to the weekly average for the entire group. By how much more or less than the weekly group average did each of the customers gain or lose? Express your answer as an integer.

The depth gauge of an unmanned submarine is tested at −500 (500 feet below the ocean's surface). If the depth gauge is incorrectly calibrated and is off by a factor of 0.1, the depth gauge error will be

$$(-500) \times (-0.1) = 50 \text{ feet}$$

This means the depth gauge will record a depth of

$$-500 + 50 = -450 \text{ feet}$$

What will the depth gauge record for each of the following calibration error factors?

29. −0.01   **30.** − 1.3   **31.** +0.05   **32.** −0.5

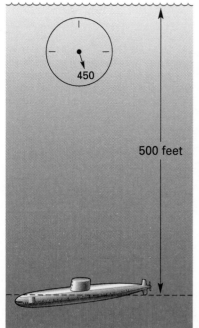

450

500 feet

LESSON 2.2 OTHER POWERS OF TEN

Negative Exponents

In Lesson 2.1, you found a pattern for writing powers of ten with positive exponents. Is there a similar pattern when the exponents are negative? A negative exponent used with base ten tells you how many times to write the factor $\frac{1}{10}$ and multiply to get the answer. That is, 10^{-4} is the same as $(\frac{1}{10})^4$.

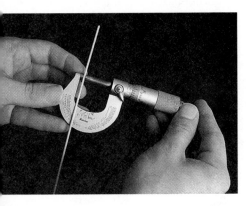

Copy the following table and try to find a pattern.

$$10^{-4} = \frac{1}{10} \times \frac{1}{10} \times \frac{1}{10} \times \frac{1}{10} = \frac{1}{10000} = 0.0001$$

$$10^{-3} = \frac{1}{10} \times \frac{1}{10} \times \frac{1}{10} = \frac{1}{1000} = 0.001$$

$$10^{-2} = \frac{1}{10} \times \frac{1}{10} = \frac{1}{100} = 0.01$$

$$10^{-1} = \frac{1}{10} = 0.1$$

How does the number of decimal places compare with the exponent?

Again, there is a pattern. The absolute value of the exponent tells you the number of times to write the factor $\frac{1}{10}$ and multiply to get the answer, as well as the number of decimal places in the answer.

Check your understanding of the table and the pattern for negative powers of ten by working through Activity 1.

ACTIVITY 1 Raising 10 to a Negative Power

1 Extend and complete the table above for 10^{-5} and 10^{-6}. Then, write a relationship giving the exponent, the number of times $\frac{1}{10}$ is used as a factor, and the number of decimal places in the answer.

2 Now do the problem "backward." First write the decimal number. Then write the power of ten it equals. The first line is done to get you started. Complete the others.

$$0.00001 = \frac{1}{10} \times \frac{1}{10} \times \frac{1}{10} \times \frac{1}{10} \times \frac{1}{10} = 10^{-5}$$

$$0.0001 = \underline{\quad\quad ? \quad\quad} = \underline{\ ?\ }$$

$$0.001 = \underline{\quad\quad ? \quad\quad} = \underline{\ ?\ }$$

$$0.01 = \underline{\quad\quad ? \quad\quad} = \underline{\ ?\ }$$

CULTURAL CONNECTION

Early civilizations used symbols for fractions. However, adding fractions using a common denominator was unknown. The early Egyptians used a method that involved rewriting each fraction as a **unit fraction sum.** Unit fractions are fractions with a 1 in the numerator.

For example, a scribe would write the fraction $\frac{2}{7}$ as $\frac{1}{4} + \frac{1}{28}$. It was the job of the scribes to make tables of as many unit fractions as the accountants and surveyors needed. But making the tables was difficult because the Egyptians wrote their numbers and letters in hieroglyphics on papyrus. Thus, the fraction $\frac{1}{4}$ was written with an ellipse as the numerator and a dot as the denominator.

How would you write $\frac{5}{6}$ using unit fractions?

$$\frac{2}{7} = \frac{1}{4} + \frac{1}{28}$$

Zero Exponents

You have written the numbers for powers of ten, such as 10^6, 10^5, $10^4 \ldots$, down to 10^{-5} and 10^{-6}. But you passed over one particular power of ten that belongs in this sequence. What does 10^0 (ten to the zero power) equal? Our previous way of writing 10 as a factor and multiplying does not work here. The statement "*write ten zero times and multiply*" does not make sense. Complete the following Activity and find a pattern.

ACTIVITY 2 | **10 to the Zero Power**

1 Copy the powers of ten table.

$$
\begin{aligned}
10^3 &= 1000. \\
10^2 &= 100. \\
10^1 &= 10. \\
10^0 &= ???? \\
10^{-1} &= 0.1 \\
10^{-2} &= 0.01 \\
10^{-3} &= 0.001
\end{aligned}
$$

2 Start at the first line. As the exponent decreases, what is happening to the number of zeros? Notice that whenever you divide a line by 10, the result is on the line below it.

3 To maintain this pattern, what is the value of 10^0?

> **Zero power of 10**
> 10^0 is another name for 1.

Critical Thinking Use a pattern to show that 2^0 is also equal to 1.

LESSON ASSESSMENT

Think and Discuss

1 Explain what each of the numbers in 3^{-2} means.

2 Explain why 10^0 and 2^0 are equal to 1.

3 How do you write the decimal representation for 10^{-3}?

4 How do you choose the exponent when writing 0.000001 as a power of ten?

5 Explain why a positive number with a negative exponent is not negative.

Practice and Problem Solving

Write each expression as an integer or a ratio without exponents.

6. 10^{-4}　　**7.** 3^{-5}　　**8.** 4^0　　**9.** 10^{-1}

10. 2^{-3}　　**11.** 5^{-2}　　**12.** 10^{-7}　　**13.** 1^{-7}

14. 10^3　　**15.** 10^{-3}　　**16.** $10^3 \times 10^{-3}$

Write each expression with a negative exponent.

17. $\frac{1}{3} \times \frac{1}{3} \times \frac{1}{3} \times \frac{1}{3}$　　**18.** $\frac{1}{5} \times \frac{1}{5} \times \frac{1}{5}$　　**19.** $\frac{1}{10} \times \frac{1}{10}$

20. The probability of rolling a 4 on a six-sided number cube is $\frac{1}{6}$. the probability of rolling a 4 twice in a row is $\frac{1}{6} \times \frac{1}{6}$. The probability of rolling a 4 three times in a row is $\frac{1}{6} \times \frac{1}{6} \times \frac{1}{6}$. What is the probability of rolling a 4 four times in a row? Express your answer as a product of fractions and as a number with a negative exponent.

21. A computer can read a number stored on its disk drive in about one nanosecond, or 10^{-9} seconds. Write this time as a ratio and as a decimal without an exponent.

Mixed Review

The lengths (in miles) of some of the principal rivers in the world are given in the table. Use < or > to compare the lengths of each pair of rivers.

Amazon	4000
Danube	1776
Ganges	1560
Mississippi	2340
Nile	4160
Ohio	1310
Rio Grande	1900
St. Lawrence	800
Volga	2194

22. Amazon; St. Lawrence

23. Volga; Mississippi

24. Danube; Ganges

25. Use an integer to show how much longer or shorter than the Rio Grande each of the rest of the rivers is.

LESSON 2.3 POWER OF TEN NOTATION

Using a Shortcut

Recall the very large number of atoms in an ounce of gold

86,700,000,000,000,000,000,000

Notice that the number 867 is followed by 20 zeros. In Lesson 2.1 you learned that 1 followed by twenty zeros is the same as writing 10^{20}. Thus, you can write

$$86,700,000,000,000,000,000,000 = 867 \times 10^{20}$$

which is now in **power of ten notation.**

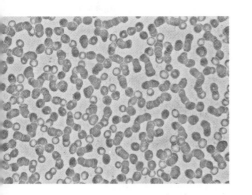

0.000007 m

A single human red blood cell is about 0.000007 meters in diameter. How can you rewrite this diameter in power of ten notation?

First write the number 7 as 7.0. Recall that $7.0 \div 10$ or $7 \times \frac{1}{10}$ is another name for 0.7. See if you can find a pattern in the following table.

$$7.0 \times \frac{1}{10} = 0.7$$
$$0.7 \times \frac{1}{10} = 0.07$$
$$0.07 \times \frac{1}{10} = 0.007$$
$$0.007 \times \frac{1}{10} = 0.0007$$
$$0.0007 \times \frac{1}{10} = 0.00007$$

Notice that multiplying a number by $\frac{1}{10}$ is the same as moving the decimal point one place to the left.

Multiplying by $\frac{1}{10}$ or dividing by 10
Multiplying a number by $\frac{1}{10}$ or dividing the number by 10 moves the decimal point one place to the left.

Think again about the number for the diameter of the red blood cell, 0.000007 meters. The previous discussion shows that you can write 0.000007 as 7 multiplied by six factors of $\frac{1}{10}$.

$$0.000007 = 7 \times \frac{1}{10} \times \frac{1}{10} \times \frac{1}{10} \times \frac{1}{10} \times \frac{1}{10} \times \frac{1}{10}$$

But you know that writing the factor $\frac{1}{10}$ six times is the same as writing 10^{-6}. So,

$$7 \times \frac{1}{10} \times \frac{1}{10} \times \frac{1}{10} \times \frac{1}{10} \times \frac{1}{10} \times \frac{1}{10} = 7 \times 10^{-6}$$

Thus, you can write the diameter of a human red blood cell as 0.000007 meters or as 7×10^{-6} meters.

The Decimal System

Since you use exactly *ten* digits (0–9) to write numbers, our system is called the **decimal system**. Each digit in a number is associated with exactly one power of ten. Thus, you can convert numbers written in decimal notation to **exponential notation**.

Decimal Notation	Expanded Notation	Exponential Notation
345	$3 \times 100 + 4 \times 10 + 5 \times 1$	$3 \times 10^2 + 4 \times 10^1 + 5 \times 10^0$
7.28	$7 \times 1 + 2 \times \frac{1}{10} + 8 \times \frac{1}{100}$	$7 \times 10^0 + 2 \times 10^{-1} + 8 \times 10^{-2}$
90.6	$9 \times 10 + 0 \times 1 + 6 \times \frac{1}{10}$	$9 \times 10^1 + 0 \times 10^0 + 6 \times 10^{-1}$

Ongoing Assessment

Write 34.06 in exponential notation.

LESSON ASSESSMENT

Think and Discuss

1 The distance from the sun to Mars is about 228 million miles. Explain how to write this distance in power of ten notation.

2 When 10 is raised to a negative power, how do you know how many places to move the decimal point? Do you move it to the left or right?

3 What happens to the decimal point when a number is multiplied by $\frac{1}{10}$?

4 Explain how to write a number in exponential notation.

Practice and Problem Solving

Write each amount in power of ten notation.

5. The Earth is about 93,000,000 miles from the sun.

6. A basketball player makes $25,000,000 on a 5-year contract.

7. There are approximately 5,500,000,000 people in the world.

8. The land area of the Earth is about 52,000,000 square miles.

9. The average diameter of a polio virus is 0.000025 millimeters.

10. It takes light about 0.000000001 seconds to travel 30 centimeters.

11. The national debt is over $4,000,000,000,000.

12. The nearest star other than the sun is about 40,000,000,000,000 kilometers away.

Write each number in exponential notation.

13. 23,683   **14.** 45.39   **15.** 4.189   **16.** 726.05

Mixed Review

Simplify.

17. $-49 \div 7$ **18.** -6×-8 **19.** $-6 + (-3)$

20. $15 - (-7) - 4$ **21.** $9 - 16 + (-12)$ **22.** $-3 \times -9 \times -2$

23. One week Roberta works 35 hours at $6.45 per hour. The same week Ricardo works 38 hours at $5.95 per hour. What is the difference in their pay for the week?

24. Business financial statements show a loss recorded in parentheses. The profit and loss from Donita's plumbing business for the first year is shown in her quarterly report. What is the total profit or loss for the year?

Q1	Q2	Q3	Q4	Total
($1246.59)	$4759.35	$8427.89	($851.17)	?

LESSON 2.4 SCIENTIFIC NOTATION

You can write the estimated world population as a power of ten.

$$5.5 \times 10^9$$

This number is also written in **scientific notation.**

> **Scientific Notation**
> A number is in **scientific notation** when it is a number greater than or equal to 1, but less than 10, multiplied by a power of ten.

Large Numbers and Scientific Notation

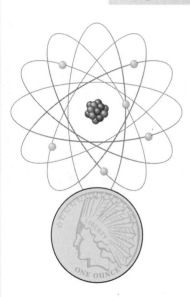

Look again at the number of atoms in an ounce of gold. In power of ten notation, you write 867×10^{20}. The number 867 is multiplied by a power of ten, but 867 is greater than 10. Thus, 867×10^{20} is not in scientific notation.

To rewrite 867×10^{20} in scientific notation, you must begin with a number that is between one and ten. Thus you must replace 867 with 8.67. But, to retain the value 867, you must multiply 8.67 by 10^2.

The result is

$$867 = 8.67 \times 10^2$$

Now 86,700,000,000,000,000,000,000 can be written in scientific notation.

$$86{,}700{,}000{,}000{,}000{,}000{,}000{,}000 = 867 \times 10^{20}$$
$$= 8.67 \times 10^2 \times 10^{20}$$

Multiplying by 10 twice, and then by 10 twenty times, is the same as using 10 as a factor 22 times. This means that

$$10^2 \times 10^{20} = 10^{22}$$

Thus,

$$86{,}700{,}000{,}000{,}000{,}000{,}000{,}000 = 8.67 \times 10^{22}$$

which is now in scientific notation.

Thus,

Critical Thinking   Explain why 5.06×10^{15} is in scientific notation.

EXAMPLE 1 Writing a Number in Scientific Notation

Write 270,000,000 molecules (the number of hemoglobin molecules in a single human red blood cell) in scientific notation.

SOLUTION

Place a pointer to the *right of the first nonzero digit*.

$$270{,}000{,}000. \text{ molecules}$$
$$\uparrow$$

Count the spaces as you move from the decimal point to the pointer ($\uparrow$). (Remember, there is a decimal point to the right of the last zero.) As you move from the decimal point to the pointer ($\uparrow$), count 8 spaces.

$$270{,}000{,}000. \text{ molecules}$$
$$\uparrow \qquad \text{8 spaces}$$

Move the decimal point to the pointer and multiply the number by 10 with an exponent equal to the number of places you moved—in this case, 8.

The result is 2.7×10^8 molecules.

Thus, 2.7×10^8 is in scientific notation because 2.7 is between 1 and 10 and it is multiplied by a power of 10.

Small Numbers and Scientific Notation

The same process works when you need to write a very small number in scientific notation.

EXAMPLE 2 Small Numbers in Scientific Notation

Write 0.000007 meters in scientific notation.

SOLUTION

Place a pointer to the right of the first digit that is not zero.

$$0.000007 \text{ meters}$$
$$\uparrow$$

Count the decimal places as you move from the decimal point to the pointer. As you move right count 6 spaces.

$$0.\underset{\text{6 spaces}}{\underbrace{000007}} \text{ meters}$$

Move the decimal point to the pointer and multiply by $\frac{1}{10}$ with an exponent equal to the number of places you moved (in this case 6). Remember that $(\frac{1}{10})^6$ is the same as 10^{-6}.

The result is 7×10^{-6} meters.

7×10^{-6} is in scientific notation because 7 is between 1 and 10 and is multiplied by a power of ten.

Ongoing Assessment

Write each number in scientific notation.

a. 93,000,000   **b.** 0.000025

Converting Scientific Notation to Decimal Notation

Sometimes it is useful to rewrite numbers in the opposite direction; that is, to convert from scientific notation to decimal notation.

EXAMPLE 3 Writing a Number in Decimal Notation

Write 1.6022×10^4 in decimal notation.

SOLUTION

Place a pointer at the decimal point.

$$1.6022 \times 10^4$$
$$\uparrow$$

Count the spaces as you move left or right from the pointer (left if the exponent is negative, right if the exponent is positive). Move the same number of spaces as the absolute value of the exponent (add zeros if needed). In this case, move right 4 spaces.

Put a decimal point after the fourth place.

16022. is in decimal notation.

Write 4.023×10^7 in decimal notation.

Numbers written in scientific notation and containing negative exponents can also be written in decimal notation.

EXAMPLE 4 Negative Exponents and Scientific Notation

Write 1.6022×10^{-19} coulombs (the charge of an electron) in decimal notation.

SOLUTION

Place a pointer at the decimal point.

$$1.6022 \times 10^{-19} \text{ coulombs}$$
$$\uparrow$$

Move left from the pointer (since the exponent is negative) 19 places (19 is the absolute value of −19), adding zeros as needed.

000000000000000000016022 coulombs

Move the decimal point to the left of the last zero you wrote and add a zero in front of the decimal point to protect it from being accidentally omitted.

0.00000000000000000016022 coulombs is in decimal notation.

LESSON ASSESSMENT

Think and Discuss

1 The average length of a bacteria is about 0.0015 centimeters. How do you write this number in scientific notation?

2 The average distance from the sun to the Earth is about 150,000,000 kilometers. How do you write this distance in scientific notation?

3 Compare power of ten notation and scientific notation. How are they alike and different?

4 When you convert a number from scientific notation to decimal notation, how do you determine how to move the decimal point?

Practice and Problem Solving

Write each amount in scientific notation.

5. 56.7   **6.** 0.0089   **7.** 245,000

8. 0.00000145   **9.** 93,673   **10.** 0.000003

11. 45,000,000   **12.** 0.00203   **13.** 135.737

14. The national debt is about 4 trillion dollars.

15. At sea level there are about 20,000,000,000,000,000,000 molecules per cubic centimeter.

16. The thickness of this page is about 0.00008 meters.

17. The wavelength of red light is about 0.0000065 meters.

Write each amount in decimal notation.

18. 4.7×10^4 **19.** 1.0×10^{-5} **20.** 7.9×10^{-7}

21. The distance light travels in one year is about 5.87×10^{12} miles.

22. The diameter of the largest moon of Jupiter is about 3.5×10^3 miles.

23. The length of a microchip is about 3×10^{-7} meters.

24. The star Alpha Centauri is about 4.1×10^{12} kilometers from Earth.

25. Alicia needs $4,675 to buy a used car. She has saved $225 for each of the last 15 months. If she continues to save at the same rate, how many more months will it take Alicia to save enough money to buy the car?

26. Julius Caesar was born in 100 BCE. He died when he was 56 years old. In what year did Caesar die?

27. The countdown clock currently shows T−85 seconds and counting. The second stage rocket will go into last check at T+20 seconds. How long after the current countdown time will the second stage check take place?

28. Your normal temperature is about 98.6°F. One morning your temperature registered 101°F. Use a positive or negative number to compare your temperature with normal.

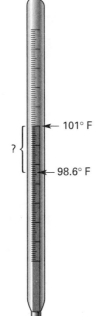

101° F

?

98.6° F

LESSON 2.5 USING SCIENTIFIC NOTATION

In Lesson 2.4, you found that multiplying $10^2 \times 10^{20}$ results in the same product as using 10 as a factor 22 times or 10^{22}. This idea leads to a rule for multiplying two numbers that have the same base but different exponents.

Multiplying Powers of Ten

10^2 means 10×10.　　　10^3 means $10 \times 10 \times 10$.

$10^2 \times 10^3$ means $10 \times 10 \times 10 \times 10 \times 10$ or 10^5

What do you notice about the exponents of the factors, 10^2 and 10^3 and the exponent in the product 10^5? Written in a different way, the problem looks like this:

$$10^2 \times 10^3 = 10^{2+3} = 10^5$$

Multiplying Powers of Ten
To multiply powers of ten, add the exponents.

Critical Thinking　Use the rule for multiplying powers of ten to show why 10^0 is equal to 1.

Dividing Powers of Ten

Is there a similar rule for dividing powers of ten?

$10^5 = 10 \times 10 \times 10 \times 10 \times 10$　　　$10^3 = 10 \times 10 \times 10$

$$\frac{10^5}{10^3} = \frac{10 \times \cancel{10} \times \cancel{10} \times \cancel{10} \times 10}{\cancel{10} \times \cancel{10} \times \cancel{10}} = 10 \times 10 = 10^2$$

Thus,

$$10^5 \div 10^3 = 10^{5-3} = 10^2$$

The same method works for $10^4 \div 10^7$.

First, rewrite the division problem as a fraction.

$$10^4 \div 10^7 = \frac{\cancel{10} \times \cancel{10} \times \cancel{10} \times \cancel{10}}{10 \times \cancel{10} \times \cancel{10} \times \cancel{10} \times \cancel{10} \times 10 \times 10} = \frac{1}{10 \times 10 \times 10}$$

$$\underset{\text{dividend} \quad \text{divisor}}{}$$

The result is

$$\frac{1}{1000} \text{ or } 10^{-3}$$

Thus, $10^4 \div 10^7 = 10^{4-7} = 10^{-3}$.

These two examples suggest a rule for dividing powers of ten.

Dividing Powers of Ten

To divide one power of 10 by another power of 10, subtract the exponent of the divisor from the exponent of the dividend.

Ongoing Assessment

Multiply or divide and write the result as a power of ten.

a. $10^{-5} \times 10^8$ **b.** $10^{-2} \times 10^{-3}$ **c.** $10^{-4} \div 10^{-6}$ **d.** $10^2 \div 10^{-5}$

Your scientific or graphics calculator uses an exponential shift key to convert, multiply, and divide numbers in scientific notation. The exponential shift key is usually designated by the EE (enter exponent) key. On some calculators you must use the 2nd key to write a number in scientific notation. Some calculators automatically convert numbers to scientific notation when they have more digits than can be displayed.

Here is how one calculator displays 256 in scientific notation.

Press the 2nd key along with the SCI key (the 5 key)
Enter 2 5 6
Press =

The display shows 2.56^{02}, which represents 2.56×10^2. Experiment with your calculator to see how it converts numbers to scientific notation.

Critical Thinking   How can you use your calculator to convert a number from scientific notation to decimal notation?

Multiplying Numbers in Scientific Notation

There are 2.7×10^8 hemoglobin molecules in a single human red blood cell. There are about 5.0×10^6 human red blood cells in one cubic millimeter (a small drop) of blood. How many hemoglobin molecules are in one cubic millimeter of blood?

To find the number of hemoglobin molecules in one cubic millimeter of blood, multiply the number of molecules by the number of red blood cells in one cubic millimeter of blood.

$$(2.7 \times 10^8) \times (5.0 \times 10^6)$$

Rewrite this problem to group the numbers between one and ten together and the powers of ten together.

$$(2.7 \times 5.0) \times (10^8 \times 10^6)$$

Use your calculator to multiply the numbers in the first parentheses. You can either multiply $10^8 \times 10^6$ mentally by adding the 8 and 6 to get 10^{14}, or you can use your calculator.

Here is one way to use a scientific calculator to find the number of molecules.

Press ②．⑦ EE ⑧
Press ✕
Press ⑤．⓪ EE ⑥
Press ═

The calculator displays 1.35^{15}. 1.35×10^{15} is the number of molecules written in scientific notation.

Ongoing Assessment

You have calculated that there are 1.35×10^{15} molecules of hemoglobin in one cubic millimeter of blood. A donation of blood to the American Red Cross consists of 5×10^5 cubic millimeters of blood. How many hemoglobin molecules were donated?

You can work a problem with negative exponents in the same way as positive exponents.

$$(2.0 \times 10^{-3}) \times (1.4 \times 10^{-2}) = (2.0 \times 1.4) \times (10^{-3} \times 10^{-2})$$

$$= 2.8 \times 10^{-5}$$

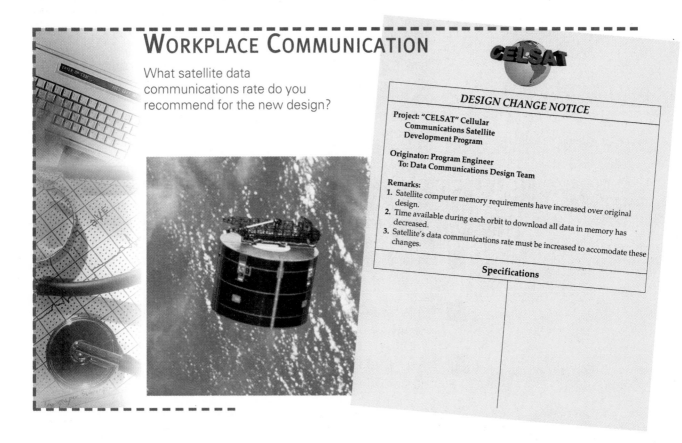

What satellite data communications rate do you recommend for the new design?

CELSAT

DESIGN CHANGE NOTICE

Project: "CELSAT" Cellular
   Communications Satellite
   Development Program

Originator: Program Engineer
To: Data Communications Design Team

Remarks:
1. Satellite computer memory requirements have increased over original design.
2. Time available during each orbit to download all data in memory has decreased.
3. Satellite's data communications rate must be increased to accomodate these changes.

Specifications

Dividing Numbers in Scientific Notation

Suppose you have a gold nugget that measures 5 millimeters on a side and has a mass of 2.41 grams. There are 7.4×10^{21} atoms of gold in the nugget. What is the mass of one atom of gold?

To find the mass of one atom, divide the mass of the nugget by the number of atoms in the nugget. First rewrite 2.41 in scientific notation as 2.41×10^0.

Now, write the division as a fraction

$$\frac{2.41 \times 10^0}{7.4 \times 10^{21}}$$

Use your calculator to simplify the fraction.

Enter 2.41
Press EE 0
Press ÷
Enter 7.4
Press EE 2 1
Press =

Thus, 3.2568×10^{-22} is the number of grams per atom.

The keystrokes used in the activity are sometimes written in the following way.

$2.41 \boxed{\text{EE}} \ 0 \div 7.4 \boxed{\text{EE}} \ 21 = 3.2568 \times 10^{-22}$

LESSON ASSESSMENT

Think and Discuss

1 Explain why you add the exponents when you multiply powers of ten.

2 Explain why you subtract the exponents when you divide powers of ten.

3 How do you multiply two numbers written in scientific notation?

4 How do you divide two numbers written in scientific notation?

5 Describe the steps used on your calculator when you multiply or divide two numbers written in scientific notation.

Practice and Problem Solving

Simplify. Write each answer as a power of ten.

6. $10^4 \times 10^{-3}$ **7.** $10^{-2} \div 10^{-4}$ **8.** $10^3 \div 10^7$

9. $10^0 \div 10^3$ **10.** $10^{-5} \times 10^9$ **11.** $10^{-7} \div 10^6$

12. $10^{-4} \times 10^4$ **13.** $10^{12} \div 10^{12}$ **14.** $10^{-3} \times 10^{-3}$

Express each product or quotient in scientific notation.

15. $(7 \times 10^5) \times (5 \times 10^3)$ **16.** $(2.6 \times 10^{-4}) \times (3.7 \times 10^6)$

17. $(5.4 \times 10^{-3}) \div (1.7 \times 10^5)$ **18.** $(7.1 \times 10^{-2}) \div (8.5 \times 10^{-7})$

19. You are setting up a laser electronics timing experiment. You need to position two mirrors so that the laser beam travels the distance between them in 6 nanoseconds (6×10^{-9} seconds). Since the speed of light is 3×10^8 meters per second, distance traveled is (3×10^8 meters per second) $\times$ (6×10^{-9} seconds).

How far apart should you position the mirrors? Express your answer in power of ten notation, scientific notation, and decimal notation.

20. In another laser experiment, you measure the speed of light in an optical fiber. The laser beam travels through 6 meters of fiber in 22 nanoseconds. You calculate

$$\text{speed of light} = (6 \text{ meters}) \div (22 \times 10^{-9} \text{ seconds})$$

Write 6 and 22×10^{-9} in scientific notation. Then compute the speed of the laser beam in the optical fiber. Write your answer in scientific notation.

Mixed Review

21. In March the normal low temperature in Denver is 25°F. The record low temperature in March is −11°F. How much greater is the normal low temperature than the record low temperature?

22. There are 8.64×10^4 seconds in one day. Write this number in decimal notation.

23. The surface temperature of the sun is about 6000 degrees Celsius. Write this number in scientific notation.

24. A special balance can weigh an object as light as 0.00000001 gram. Write this number in scientific notation.

Measurement is an important part of many jobs. When you measure the dimensions of an object, you are comparing the dimensions to a standard unit of measure. On most jobs, you will measure length, mass, or capacity. These measurements are usually made in feet, pounds, or gallons. But in science and industry, these measurements are more often made using the basic standards of the metric system: meters, grams, and liters.

The metric system is a decimal system. Exponents of ten and scientific notation make it easy to convert measurements within the metric system. Prefixes used in the metric system are shown in the following table.

Value	Exponent	Symbol	Prefix
1 000 000 000 000	10^{12}	T	tera
1 000 000 000	10^{9}	G	giga
1 000 000	10^{6}	M	mega
1 000	10^{3}	k	kilo
100	10^{2}	h	hecto
10	10^{1}	da	deca
0.1	10^{-1}	d	deci
0.01	10^{-2}	c	centi
0.001	10^{-3}	m	milli
0.000001	10^{-6}	μ	micro
0.000000001	10^{-9}	n	nano
0.000000000001	10^{-12}	p	pico

You can write these prefixes before any metric unit of measure. For, example, they can precede the meter, the gram, or the liter. Although the meter is the basic unit for measuring length, the centimeter and the kilometer are used to measure smaller or larger distances.

You also use metric prefixes for units of time such as seconds. The data in computers move at speeds measured in nanoseconds (10^{-9} seconds) and even picoseconds (10^{-12} seconds). The prefixes are used for many electrical units, such as ohms, amperes, volts, and watts. A megaohm is a million ohms. A microamp is one one-millionth of an ampere.

The following table helps you convert from one metric unit to another. Notice that the basic units (meter, gram, and liter) are in the middle box of the first column and in the middle box of the top row.

TO CONVERT		TO MILLI-	TO CENTI-	TO DECI-	TO METER GRAM LITER	TO DECA-	TO HECTO-	TO KILO-
	KILO-	$\times$ 10^6	$\times$ 10^5	$\times$ 10^4	$\times$ 10^3	$\times$ 10^2	$\times$ 10^1	
	HECTO-	$\times$ 10^5	$\times$ 10^4	$\times$ 10^3	$\times$ 10^2	$\times$ 10^1		$\times$ 10^{-1}
	DECA-	$\times$ 10^4	$\times$ 10^3	$\times$ 10^2	$\times$ 10^1		$\times$ 10^{-1}	$\times$ 10^{-2}
	METER GRAM LITER	$\times$ 10^3	$\times$ 10^2	$\times$ 10^1		$\times$ 10^{-1}	$\times$ 10^{-2}	$\times$ 10^{-3}
	DECI-	$\times$ 10^2	$\times$ 10^1		$\times$ 10^{-1}	$\times$ 10^{-2}	$\times$ 10^{-3}	$\times$ 10^{-4}
	CENTI-	$\times$ 10^1		$\times$ 10^{-1}	$\times$ 10^{-2}	$\times$ 10^{-3}	$\times$ 10^{-4}	$\times$ 10^{-5}
	MILLI-		$\times$ 10^{-1}	$\times$ 10^{-2}	$\times$ 10^{-3}	$\times$ 10^{-4}	$\times$ 10^{-5}	$\times$ 10^{-6}

Follow these steps to use the chart.

1. Find the name in the column at the left for the unit you are converting.

2. Go across this row to the column with the heading of the unit you are converting to.

3. Multiply the unit you are converting by the power of ten shown in the box.

Do the following Activity to see how to use the chart to convert metric measures.

Converting Metric Measures

How do you change 35 milliamps to amperes? An ampere is another basic unit like the meter, gram, or liter.

TO CONVERT

	TO MILLI-	TO CENTI-	TO DECI-	TO METER GRAM LITER	TO DECA-	TO HECTO-	TO KILO-
KILO-	$\times 10^6$	$\times 10^5$	$\times 10^4$	$\times 10^3$	$\times 10^2$	$\times 10^1$	
HECTO-	$\times 10^5$	$\times 10^4$	$\times 10^3$	$\times 10^2$	$\times 10^1$		$\times 10^{-1}$
DECA-	$\times 10^4$	$\times 10^3$	$\times 10^2$	$\times 10^1$		$\times 10^{-1}$	$\times 10^{-2}$
METER GRAM LITER	$\times 10^3$	$\times 10^2$	$\times 10^1$		$\times 10^{-1}$	$\times 10^{-2}$	$\times 10^{-3}$
DECI-	$\times 10^2$	$\times 10^1$		$\times 10^{-1}$	$\times 10^{-2}$	$\times 10^{-3}$	$\times 10^{-4}$
CENTI-	$\times 10^1$		$\times 10^{-1}$	$\times 10^{-2}$	$\times 10^{-3}$	$\times 10^{-4}$	$\times 10^{-5}$
MILLI-		$\times 10^{-1}$	$\times 10^{-2}$	$\times 10^{-3}$	$\times 10^{-4}$	$\times 10^{-5}$	$\times 10^{-6}$

1 Since you are converting milliamps, find "milli" in the column at the left.

2 Since you are converting to amperes, move across the row until you reach the column with the basic units for the heading. Since ampere is a basic metric unit, you can add ampere (or any other standard metric unit) to the top center box of the chart. The power of ten in the box is 10^{-3}.

3 Multiply 35 by 10^{-3}.

$$35 \text{ milliamps} = 35 \times 10^{-3} \text{ amperes}$$

Remember, multiplying by a negative power of ten moves the decimal point to the left as many places as the absolute value of the exponent. Thus 35×10^{-3} amperes is the same as 0.035 amperes.

Critical Thinking Change 4.5 meters to centimeters. What happens to the decimal point when you multiply 4.5 by 10^2?

LESSON ASSESSMENT

Think and Discuss

1 Why are standard units of measure needed?

2 Explain how to use the metric conversion table to convert centimeters to meters.

3 Explain how to use the metric conversion table to convert kilograms to milligrams.

4 Why is the metric system rather than the English system preferred by scientists and engineers?

Practice and Problem Solving

Use the metric conversion table on page 29 to change

5. 34 meters to kilometers.

6. 6.8 milliliters to liters.

7. 5.8 centimeters to millimeters.   .

8. 17 kilowatts to watts.

9. 350 milliseconds to seconds.

10. 6 millimeters to centimeters.

11. 0.8 kilograms to milligrams.

12. 0.006 meters to millimeters.

13. 8.3 centimeters to meters.

14. 1000 grams to kilograms.

15. 142 milliamps to amps.

16. 500 liters to milliliters.

17. What power of ten is used to convert picoseconds to seconds?

18. What power of ten is used to convert gigabytes to bytes?

The windchill is determined from air temperature and wind speed.

Wind Speed in Miles Per Hour	Air Temperature in °F			
0	20	10	0	−10
10	3	−9	−22	−34
20	−10	−24	−39	−53
30	−18	−33	−49	−64

Use the windchill table to find each of the following.

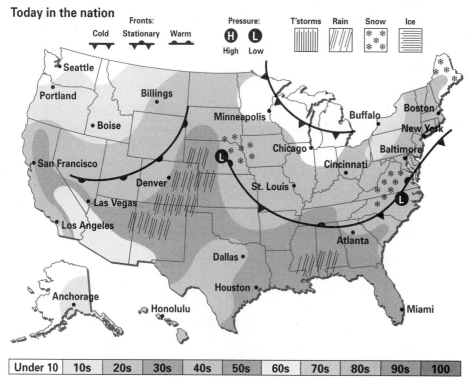

Today in the nation

19. In Boston the air temperature is 20°F and the wind is blowing 20 miles per hour. What is the windchill?

20. In Chicago the windchill is reported to be −22 with a 10 mile per hour wind. What is the air temperature?

21. The air temperature in Buffalo is 10°F. The wind suddenly increases from 10 to 20 miles per hour. Use an integer to describe the change in windchill.

22. Suppose your watch loses 1.5 minutes per day. Use integers to describe the total minutes lost after 10 days?

23. The diameter of the sun is about 1,392,000 kilometers. Write the diameter in scientific notation.

Perform each operation. Give your answer in scientific notation.

24. $(8 \times 10^4) \times (1.4 \times 10^{-6})$ **25.** $(1.458 \times 10^{-5}) \div (2.7 \times 10^9)$

MATH LAB

Activity 1: Measuring Average Paper Thickness

Equipment   Calculator
Vernier caliper
Micrometer caliper
Four large books of 500 or more numbered pages

Problem Statement

You will find the average paper thickness of four books. You will then compare your answer to the paper thickness measured with a micrometer caliper.

Procedure

a Use the vernier caliper to measure the total thickness (to the nearest thousandth of an inch) of pages 1-500 for each book. Record the thickness under these headings:

Book A   Book B   Book C   Book D

b Calculate an average paper thickness for each book by dividing the measured thickness by 250 (there are two numbered pages for each sheet of paper). Write your group's answer for the average page thickness of each book in scientific notation. Record the results for each group.

c Calculate the class average of the paper thickness for each book. Write each class-averaged paper thickness in scientific notation.

d How does the average paper thickness calculated by your group compare to the class-averaged paper thickness for each book? Which is more accurate?

e Use a micrometer caliper to measure the thickness of a single sheet of paper in each book. Record your answer in scientific notation. How does the paper thickness measured directly with the micrometer caliper compare with each paper thickness your group calculated? Which is more accurate?

Activity 2: Counting Sand Grains

Equipment   Calculator
Graduated cylinder
Micrometer caliper
Coarse sand such as kitty litter

Problem Statement

You will count the number of grains of sand in a small volume and use this number to estimate the number of grains in a larger volume. You will also estimate the volume of an average grain of sand.

Procedure

a Use the graduated cylinder to measure 1 milliliter of sand. Count the grains of sand in 1 milliliter.

b From your count, how many grains of sand would you estimate are in a cubic meter? (1 milliliter is equal to 1 cubic centimeter.) Write your answer in scientific notation.

c From your count, how many grains of sand would you estimate are on a strip of beach 1500 meters long by 200 meters wide down to a depth of 10 centimeters? Write your answer in scientific notation.

d From your answer to **a,** what is the average volume (in cubic centimeters) of one grain of sand? Write your answer in scientific notation.

e Isolate 4 or 5 grains of sand and measure their diameter with a micrometer caliper. From these measurements, calculate an average diameter.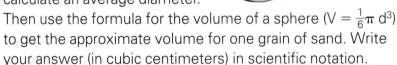
Then use the formula for the volume of a sphere ($V = \frac{1}{6}\pi d^3$) to get the approximate volume for one grain of sand. Write your answer (in cubic centimeters) in scientific notation.

f How do the values for **d** and **e** compare?

Equipment Calculator
String
Spring scale with 5000-gram capacity
Graduated cylinder, 500-milliliter capacity
Water supply and drain

Problem Statement

You will measure the volume and weight of a sample of water. You will use this data to calculate the number of molecules in the sample and the average volume and weight of each molecule.

Procedure

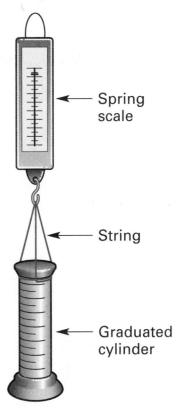

a Tie a string around the top of the graduated cylinder (see illustration). Weigh the empty graduated cylinder and string by hooking the loop on the spring scale. Record this weight.

b Measure 500-milliliters of water in the graduated cylinder. Use the spring scale to find the combined weight of the graduated cylinder and water. Record the combined weight.

c Subtract the weight of the empty graduated cylinder and string from the combined weight of the water and graduated cylinder. This is the weight of the water sample. There are 6.02×10^{23} molecules of water in 18 grams of water. Determine the number of molecules in your 500-milliliter sample. Write your answer in scientific notation.

Spring scale

String

Graduated cylinder

d Based on the data in **c,** calculate the average volume per molecule and the average weight per molecule. Write your answer in scientific notation.

MATH APPLICATIONS

The applications that follow are like the ones you will encounter in many workplaces. Use the mathematics you have learned in this chapter to solve the problems. Wherever possible, use your calculator to solve the problems that require numerical answers. A table of conversion factors can be found beginning on page A1.

A tropical year is 365.24220 days. A year expressed on a Gregorian calendar is 365.2425 days.

1 What is the difference between a tropical year and the Gregorian calendar year? Express your answer in scientific notation.

JANUARY	FEBRUARY					MARCH
Sunday	Monday	Tuesday	Wednesday	Thursday	Friday	Saturday
			1	2	3	4
5	6	7	8	9	10	11
12	13	14	15	16	17	18
19	20	21	22	23	24	25
26	27	28	29	29.0003		

2 How many years will it take for this difference to amount to one day?

3 Some home heaters emit pollutants into the air. The maximum acceptable level of carbon monoxide in a room is about 9 parts per million. For nitrogen dioxide a value of about 0.05 parts per million is the suggested amount. Express the maximum acceptable concentration of each gas as a percentage written in scientific notation.

A laser printer prints characters at a density of 12 dots per millimeter on a piece of paper 216 millimeters wide and 279 millimeters long.

4 How many dots can be printed across the paper?

5 How many dots can be printed down the paper?

6 How many dots can be printed on the surface of the paper? Express your answer in scientific notation.

7 If each dot just touched an adjacent dot, what is the diameter of a dot? Express your answer in scientific notation.

The nearest star is about 4.5 light years away. A light year is the distance light can travel in one year. The speed of light is approximately 186,000 miles per second.

8 Convert the speed of light to miles per year by multiplying the speed by the number of seconds in a year. This is the distance light travels in a year.

9 How far away (in miles) is the nearest star? Express your answer in scientific notation.

A recent report of leading corn-growing states in the United States shows the following annual amounts.

AGRICULTURE & AGRIBUSINESS

State	Annual Corn Growth (Bushels)
Iowa	1,739,900,000
Illinois	1,452,540,000
Nebraska	802,700,000
Minnesota	744,700,000
Indiana	654,000,000
Wisconsin	378,000,000
Ohio	360,000,000
Michigan	273,600,000
Missouri	213,400,000
South Dakota	180,000,000

10 What is the total number of bushels of corn grown by these states? Express your answer in scientific notation.

11 A bushel of corn weighs about 56 pounds. How many pounds of corn did these states produce?

12 The total annual production was reported to be 8,201,000,000 bushels. How many bushels were produced by the states not listed? Express your answer in scientific notation.

When diesel fuel burns, it releases about 138,400 Btu of energy per gallon. A tractor uses 215 gallons of diesel fuel during 56 hours of operation.

13 How many Btu of energy are released by the 215 gallons of fuel? Express your answer in scientific notation, rounded to two decimal places.

14 Divide the total energy released by 56 hours to find the rate of energy released in Btu per hour. A rate of energy release of 2546 Btu per hour is equivalent to 1 horsepower. Find the number of horsepower released by the diesel fuel over 56 hours of operation.

15 The tractor is doing work at a rate of 70 horsepower. This rate of work is what percent of the horsepower released by the fuel? This is the "thermal efficiency" of the engine.

In one region of Hong Kong, 55,000 persons live on about 24 acres.

16 Convert the 24 acres to square feet. Express your answer in scientific notation.

17 On the average, how many people live on each square foot of space in this region?

18 Multiply the answer to Exercise 17 by the number of square miles per square foot to determine about how many people are living in each square mile in this region.

19 Repeat the above calculations using the population for your city or town, and the area of your town. How does the population per square mile for Hong Kong compare to where you live?

BUSINESS & MARKETING

You are making a bid to operate the concession stands at a play-off game. You expect a very large crowd, possibly 85,000 fans. Past experience has shown that each fan will purchase an average of about 1.75 cups of soft drink.

20 How many cups of soft drink can you expect to sell? Express your answer in scientific notation.

21 If each cup is filled with 6 ounces of ice, how many ounces of ice can you expect to use? Express your answer in scientific notation.

A newspaper publisher uses 0.17 pounds of ink for every 1000 pages of print. The newspaper averages 80 pages per day. The distribution averages about 130,000 papers per day.

22 What is the average amount of ink on each page of newsprint? Express your answer in scientific notation.

23 What is the approximate total number of pages of newsprint produced during a week? Express your answer in scientific notation.

24 About how much ink is used during a week's printing? Express your answer in both scientific notation and decimal notation.

A report shows that annual consumer spending for professional services rose from $260,000,000,000 to $410,000,000,000 over a 10-year period.

25 Express each amount in scientific notation.

26 What is the growth in spending for services each year? Express your answer in scientific notation.

A newspaper reports that the average price of a new home in the United States is $93,000. The report estimates that 65,000 new homes were sold nationwide during a recent period.

27 Approximately how much money was spent on new homes during this period? Express your answer in scientific notation.

28 A friend states that "megabucks" are spent on new homes. Using your knowledge of prefixes and powers of ten, is your friend right to use the term "megabucks"? If so, how many "megabucks" are being spent?

29 The subcompact division of Luxury Motors has produced a total of 17,400,000 cars at its 3 assembly plants. Reports of dissatisfaction with the steering wheel design of these models have been reported from 281 owners. What percent of the owners have expressed dissatisfaction with the steering wheel? Express your answer in scientific notation.

A computer printout has been "formatted" to report its values in scientific notation.

Beginning inventory	7.125 E + 04
Additional purchases	2.450 E + 03
Production adds	1.035 E + 05

30 What is the total of the three figures reported?

31 The data were rounded off for the report. Write each of the three numbers in decimal notation and indicate where you think the rounding occurred in each number.

32 Suppose that a laser technician wanted to see more digits (less rounding). She suggests reporting the inventory in terms of "thousand units." Rewrite the data in terms of "thousand units."

Suppose the national budget is $1.8 trillion ($1,800,000,000,000).

33 Express the budget figure in scientific notation.

34 If the population of the country is 250,000,000, how much is budgeted per person in the country?

A report states that credit card users charged a total of $150 billion during a year. During the reported time period, lenders collected a total of $12.6 billion in interest on outstanding charges.

35 Express each of the dollar figures in scientific notation.

36 What percent of the charged amount is the interest amount collected by the lenders?

A computer sales brochure advertises that its new computer system has a 16-megabyte memory and can store a total of nearly a half-gigabyte of data on its three hard disk drives.

37 If a single character occupies a byte of memory or disk space, how many characters can be stored in the 16-megabyte memory? Express your answer in scientific notation and decimal notation.

38 How many characters can be stored on each of the three disk drives? Express your answer in scientific notation and decimal notation.

39 A warehouse has 7 rooms. Each room has 8 rows of 120 bays. Each bay holds 25 pallets. Each pallet contains 40 tires. Each tire averages 35 pounds. Approximately how many pounds of tires can the warehouse hold? Express your answer in scientific notation.

A synthetic thyroid preparation that is available in 25-microgram amounts for each tablet is prescribed to a patient. The bottle of medication contains 30 tablets.

40 How many grams of medication are in each tablet? Express your answer in scientific notation.

41 How many grams of medication are in each bottle of medication? Express your answer in scientific notation.

A group of researchers conducted independent studies of a bacterium. The researchers recorded the average length of the bacteria.

Researcher	Average Reported Length
Choy	0.001633 micrometers
Robert	1.542 nanometers
Salsido	0.001491 micrometers
Dana	1584 picometers

42 Express each measurement in scientific notation in units of meters.

43 Determine the average of the four researchers' results. Express your answer in scientific notation.

44 What is the difference between the largest and smallest reported result? (This is called the "range" of the reported results.)

45 Suppose a particular bacterium can divide into two bacteria, once each hour. Thus, in one hour, one bacterium becomes two. After two hours, two bacteria become four (2 × 2). In three hours, they become eight (2 × 2 × 2), and so on. If this process continues for a total of 48 hours, how many bacteria will there be? Write this number in scientific notation and in decimal notation.

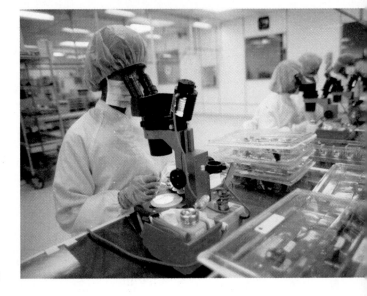

Three samples of ocean salt are weighed on an electronic balance. The samples weigh 0.052476 grams, 0.052482 grams, and 0.052490 grams.

46 Express each weight in scientific notation.

47 What is the difference in weight between the largest and smallest samples? Express your answer in scientific notation.

48 What is the prefix that denotes the smallest increment of weight reported using this balance (the last digit)?

INDUSTRIAL TECHNOLOGY

The spark plugs in most automotive engines must provide a spark for every two revolutions of the engine. When a car is driven at normal speeds, the engine turns at about 2400 revolutions per minute (rpm).

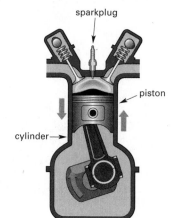

sparkplug

piston

cylinder→

49 If you assume an average speed of 1 mile per minute and an annual mileage of 10,000 miles, approximately how many minutes is the car driven during the year?

50 At 2400 revolutions per minute, or 1200 sparks per minute, during the year's driving, about how many times does the spark plug fire during the year? Express your answer in scientific notation.

51 Cosmic rays entering the earth's atmosphere from outer space typically have an energy of about 2 GeV, or 2 giga-electron volts. Express the energy of these rays in electron volts (eV) using scientific notation.

A thermocouple made with lead and gold wires is designed to measure temperature. The thermocouple produces an electric potential of 2.90 microvolts for each Celsius degree above 0°C.

52 Express the thermocouple output as volts per Celsius degree in scientific notation.

53 What voltage would you expect from a thermocouple at 15°C? Express your answer in scientific notation.

A certain isotope of radioactive uranium has a half-life of 4.5×10^9 years. This means that in 4.5×10^9 years, a sample of this isotope will have one-half the level of radioactivity it has today.

54 What is the half-life of this isotope written in decimal notation?

55 If a generation is considered to be 40 years, how many generations will it take for this isotope's radioactivity to be reduced by half?

Radio and television frequencies are given in hertz, or cycles per second. Listed below are frequencies for other common forms of electromagnetism. Use the prefixes to convert each frequency to scientific notation in units of Hz.

Type of Broadcast	Frequency
56 Household electricity	60 Hz
57 AM radio	1080 kHz (kilohertz)
58 Shortwave radio	10 MHz (megahertz)
59 FM radio	102 MHz (megahertz)
60 Radar	8 GHz (gigahertz)
61 Microwave communication	12 GHz (gigahertz)
62 Visible light	400 THz (terahertz)

A quartz crystal is commonly used in a digital watch to provide a stable time base. The crystal oscillates with a frequency of 5.0 megahertz. One hertz is equivalent to one vibration or cycle per second.

63 Express the quartz frequency of vibration in scientific notation, with units of hertz.

64 With this frequency, very small time divisions are possible. The smallest time division is the time for one cycle of the crystal vibration. This cycle period is computed by dividing 1 by the frequency. What is the smallest time division for this quartz crystal in seconds? Express your answer in scientific notation.

A commonly measured value of chemical solutions is pH. This is a measure of hydrogen-ion activity. The hydrogen-ion activity of pure water at 25°C is 0.0000001 mole per liter. A highly acidic solution has 1.0 mole per liter of hydrogen-ion activity, while a highly basic solution has an activity of 0.00000000000001 mole per liter.

65 Express each of the three hydrogen-ion activities in scientific notation.

66 The pH value of a solution is simply the absolute value of the exponent of ten in the measure of its hydrogen-ion activity. What is the pH value of the highly acidic solution above, of pure water, and of the highly basic solution?

A computer is advertised as having a processing speed of "11 mips," or 11 million instructions per second.

67 Express this speed in scientific notation.

68 How long does it take to process each instruction at such a speed?

69 How many "nanoseconds" is this?

Steel can be stretched when a stress is applied. The "modulus of elasticity" for steel is 30×10^6 pounds per square inch (psi). The elongation (or lengthening because of a stress) of a steel beam can be calculated by multiplying the length of the beam by the stress (in psi) and dividing by the modulus of elasticity.

70 Convert the length of a 12-foot steel beam to inches.

71 Suppose a stress of 5000 psi is applied to the beam. Compute the elongation of the beam. Express your answer in scientific notation.

A furnace used to process aluminum consumes approximately 25,000,000 watt-hours of energy per ton (2000 lb) of aluminum processed.

72 Express the energy consumed in scientific notation.

73 If a furnace processes 3200 tons of aluminum during a given period, about how much energy is used? Express your answer in scientific notation.

74 Energy is usually reported in kilowatt-hours (kWh) rather than watt-hours. Convert the answer to Exercise 73 to kWh.

An elevator with a mass of 1200 kg is lifted upward at a constant speed of 2.0 meters per second.

75 Multiply the mass of the elevator by 9.8 newtons per kg to find the force needed to lift the elevator. Express your answer in scientific notation.

76 Multiply the force by the speed to find the power (in watts) needed to lift the elevator.

77 Power is commonly reported in kilowatts (kW). How many kilowatts of power are needed to lift this elevator?

Electricity travels close to the speed of light, 3.0×10^8 meters per second. A surge suppressor used to protect computer equipment advertises a 1-ns clamping time (that is, it can suppress a voltage pulse in 1 nanosecond).

78 Express the advertised clamping time in units of seconds using scientific notation.

79 Multiply the speed of the electricity in a wire by the clamping time to estimate how far along the wire (in meters) the leading edge of a voltage pulse will travel before the remainder of the pulse is suppressed.

Security coding of automatic garage door openers is done by setting a bank of switches. For example, with three such switches that can be set to ON or OFF, the number of possible settings is $2 \times 2 \times 2$, or 8 settings. Some switches have three possible settings, indicated as +1, 0, or −1, yielding a total of $3 \times 3 \times 3$, or 27 possible settings for three switches.

80 How many possible settings are in a switch bank of 8 switches with three settings for each? Express your answer in scientific notation.

81 How many possible settings are in a bank of 12 switches with three settings for each? Express your answer in scientific notation.

Skills

Write each number as a power of ten.

1. 1000   **2.** 10,000,000   **3.** 0.00001

Write each number in decimal form.

4. 1.1×10^2 **5.** 4.5×10^{-6} **6.** 1.89×10^4

Write each number in scientific notation.

7. 0.00068   **8.** 25,000,000   **9.** 1,045,000,000

Write each number in decimal form.

10. 3^4 **11.** 5^{-3} **12.** $(-2)^5$

Express each product or quotient in scientific notation.

13. $(8.3 \times 10^{-2}) \times (1.3 \times 10^8)$ **14.** $(7.5 \times 10^5) \div (1.5 \times 10^{-3})$

Convert the metric units and write each answer in scientific notation.

15. 75 centimeters to kilometers.

16. 432 seconds to microseconds.

Applications

Express each answer in scientific notation.

17. If 6.02×10^{23} molecules of gas have a mass of 28 grams, what is the mass of one molecule of the gas?

18. If 1 cubic centimeter of water has a mass of 1 gram, what is the volume in cubic centimeters of 2538 kilograms of water?

19. One gallon of water has a mass of 3629 grams. If 18 grams of water contain 6.02×10^{23} molecules, how many molecules are in 1 gallon of water?

20. One cubic centimeter of water has a mass of 1 gram. A cubic tank is 8 meters on each edge. How many grams of water does the tank hold?

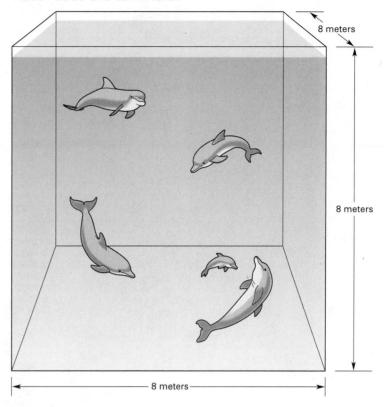

8 meters

8 meters

8 meters

Math Lab

21. Which measuring instrument is the most accurate for finding the average page thickness in a book?

a. micrometer caliper
b. vernier caliper
c. machinist ruler

22. The average volume of a sand grain on a beach is 0.01 cubic centimeters. Estimate the number of grains of sand in a section of beach that is 1000 meters long, 100 meters wide, and 1 meter deep. Express your answer in scientific notation.

23. In the math lab "Measuring Water Molecules", why was the weight of the empty graduated cylinder and string measured?

CHAPTER 3

WHY SHOULD I LEARN THIS?

Formulas are the foundation of technology—they describe our world and how it works. Learn how to build formulas that you can use in your future career.

USING FORMULAS

OBJECTIVES

1. Substitute and evaluate values in expressions.
2. Read and write a formula.
3. Use your calculator to solve problems with formulas.

A formula is used to show how quantities are related to one another.

Although you have used many formulas before, you may not have stopped to think about what a formula does or how one is created. A formula in mathematics is a way of using symbols to write a sentence.

Sentence: The perimeter of a rectangle is the sum of twice its length plus twice its width.

Formula: $p = 2l + 2w$

Do you think it is easier to remember the sentence or the formula?

Nearly every occupation uses some kind of formula. Here are some examples.

Carpenters use formulas to check right angles.
Hospital personnel use formulas to decide dosage.
Bricklayers use formulas to mix mortar.
Dietitians use formulas to balance nutrition amounts.
Environmentalists use formulas to determine concentrations.
Photographers use formulas to mix chemicals
Potters use formulas to create glazes.

As you watch the video for this chapter, notice how often you see people using formulas to do their jobs.

LESSON 3.1 VARIABLES AND EXPRESSIONS

A **numerical expression** contains addition, subtraction, multiplication, or division operations. In the first two chapters, you evaluated many simple numerical expressions, such as 3^2 and 6×5. The expression 6×5 can also be written with a raised center dot or parentheses to indicate multiplication.

$$6 \cdot 5 = 30 \qquad 6(5) = 30 \qquad (6)5 = 30$$

How do you find the value of a numerical expression that contains several different operations?

Order of Operations

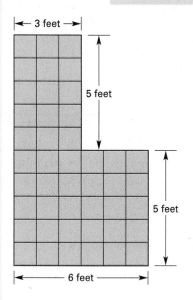

A flooring contractor needs to tile two rectangular areas. The dimensions of the two areas are 6 feet by 5 feet and 3 feet by 5 feet. How many square feet of tile should the contractor order?

To find the total number of square feet, the contractor must find the value of $6 \cdot 5 + 3 \cdot 5$. Which operation, addition or multiplication, is performed first?

If you add first and then multiply, the value is $6 \cdot 8 \cdot 5$ or 240.

If you multiply first and then add, the value is $30 + 15$ or 45.

To make sure everyone gets the same value when they evaluate the same numerical expression, mathematicians have developed the following **order of operations.**

> ### The Order of Operations
> First, perform all operations within parentheses.
>
> Second, perform all operations involving exponents.
>
> Third, multiply or divide in order from left to right.
>
> Finally, add or subtract in order from left to right.

When the flooring contractor follows the order of operations, his result is

$$6 \cdot 5 + 3 \cdot 5 = 30 + 15$$
$$= 45$$

The contractor should order enough tile to cover 45 square feet.

If you want to perform the addition first, use parentheses.

$$6 \cdot (5 + 3) \cdot 5 = 6 \cdot 8 \cdot 5$$

$$= 240$$

Evaluate each expression.

a. $9 \cdot 2 - 24 \div -6$ **b.** $5 + 2(4 - 7)^2$

Scientific calculators usually follow the order of operations. Use your calculator to evaluate $6 \cdot 5 + 3 \cdot 5$. If your calculator does not display 45, you will need to use parentheses or the memory key.

Algebraic Expressions

Every time you hire an electrician, it costs $50 for the house call. The total bill, however, will depend on the number of hours the electrician is on the job. A table is useful in finding the cost for the total number of hours on the job when the hourly cost is $30.

Hours	Expression	Cost
1	$50 + 30 \cdot 1$	80
2	$50 + 30 \cdot 2$	110
3	$50 + 30 \cdot 3$	140

The table shows that the cost depends on the number of hours the electrician works on the job. How can you find the cost after any number of hours on the job? If you think of the letter h as representing the number of hours, the cost is

$$50 + 30 \bullet \text{number of hours } (h)$$

or simply, $50 + 30 \bullet h$.

In the expression $50 + 30 \bullet h$, the letter h is a **variable.** The numbers 50 and 30 are **constants.**

A **variable** is a symbol, usually a letter, used to represent a number or some other object. However, a letter that always represents the same number is a constant. For example, the Greek letter π (pi) always represents the same number (approximately 3.1416). Thus, π is a constant.

An expression containing numbers, operations, and one or more variables is called an **algebraic expression.** Each of the following expressions represents the multiplication of a and b.

$$ab \qquad a(b) \qquad (a)b \qquad (a)(b)$$

In the algebraic expression representing the electrician's cost,

$$50 + 30h$$

50 and $30h$ are the **terms** of the expression. In the term $30h$, 30 is the numerical coefficient, or just the **coefficient** of h.

Ongoing Assessment

Practice the language of algebra by answering the following questions:

a. In the expression $ab + ac$, which operation is performed first?

b. What is the value of $5n$ if $n = 7$?

c. Name the variable in $5n$.

d. Name the constant in $5n$.

e. Name the coefficient in $5n$.

To evaluate an algebraic expression, first substitute the given numbers for the variables. Then simplify the resulting numerical expression using the order of operations.

Evaluate the expression $5x - y^2$ when x is -2 and y is 3.

SOLUTION

1. Substitute the numbers for the variables.

$$5x - y^2$$

$$5(-2) - (3)^2$$

2. Follow the order of operations.

$$5(-2) - 9$$

$$-10 - 9$$

$$-19$$

Critical Thinking   For what value of b, will $25 - b^2 = 9$?

LESSON ASSESSMENT

Think and Discuss

1 Why is it important to have an order of operations?

2 Compare the meanings of a variable, a constant, and a coefficient. Give an example of each.

3 What is the difference between a numerical expression and an algebraic expression?

4 Explain how to evaluate an algebraic expression.

Practice and Problem Solving

Evaluate each numerical expression.

5. $25 + 4 \bullet 10$ **6.** $9 \div 3 - 15 \div 5$ **7.** $4 - (9 - 12)$

8. $8 - (7 - 9)^3$ **9.** $36 \div 6 \div 3$ **10.** $36 \div (6 \div 3)$

Evaluate each algebraic expression.

11. $6a - 10$, when a is -3 **12.** $100 - 5c$, when c is 100

13. $-5(x + 6)$, when x is 14 **14.** $8y - 3y$ when y is -2

15. $2m + 2n$, when m is 6 and n is 4

16. $s^3 - t^2$, when s is -5 and t is -10

Evaluate each expression when r is -6, s is 4, and t is -2.

17. $rs - t$ **18.** $r \div t + st$ **19.** rst

20. $rs - rt$ **21.** $(r + s)^2$ **22.** $r^2 + s^2$

23. A stockbroker has developed the expression $P \div (G + Y)$ to help her determine if the price of a given stock is too high. Evaluate the expression when P is 20.25, G is 4.50, and Y is 2.25.

24. The expression $P(0.08L + I)$ helps a clothing store manager estimate projected sales for a given item. Evaluate the expression when P is 20, L is 350, and I is 45.

25. The manufacturing cost in dollars for a type of wooden shelf is found using the expression $br + ac$. Find the manufacturing costs when b is 4, r is 8.5, a is 48, and c is 1.5.

Mixed Review

26. The Johnson Brothers Company has lost $240 for 6 straight weeks. Use integers to write an expression to represent the total losses. Simplify the expression.

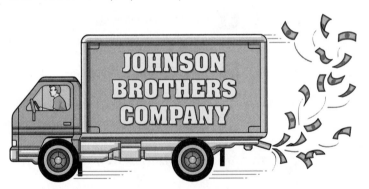

27. The estimated age of the earth is 4.7×10^9 years. Write the age of the earth as in decimal notation.

Simplify and write each answer in scientific notation.
28. $(3 \times 10^3) \cdot (6 \times 10^{-5})$ **29.** $(3.6 \times 10^2) \div (1.2 \times 10^4)$

LESSON 3.2 EQUATIONS AND FORMULAS

In Lesson 3.1, you evaluated algebraic expressions for given values of a variable. Complete Activity 1 and see if you can find a pattern.

ACTIVITY 1   Equal Expressions

Evaluate each pair of expressions. Remember to use the order of operations. Symbols such as parentheses ( ) and brackets [ ] are called **grouping symbols.** Do the operations in the innermost set of grouping symbols first.

Column 1	Column 2
1 $3 \cdot 4 + 3 \cdot 6$	$3(4 + 6)$
2 $(8 \cdot -7) + (8 \cdot -3)$	$8[-7 + (-3)]$
3 $6 \cdot 28 + (6 \cdot -8)$	$6[28 + (-8)]$

4   What do you notice about the answer to each pair of expressions?

Equations and Basic Properties

An **equation** is a sentence that states two expressions are equal. Equations are used to model the data in all kinds of situations.

ACTIVITY 2   The Distributive Property

1   A carpet installer is installing two rectangular carpets, one measuring 3 feet by 4 feet and the other measuring 3 feet by 6 feet. To find the total carpet area for this job, simplify the following expression:

$$3 \cdot 4 + 3 \cdot 6$$

2 On another job, he needs to carpet a 3 foot by 10 foot area. The area for the second job is 3 • 10 or

$$3(4 + 6)$$

3 Are the areas of the two jobs equal?

The equation
$$3 • 4 + 3 • 6 = 3(4 + 6)$$

models the carpet installer's situation. The pattern represented by this equation can be written with variables.

$$ab + ac = a(b + c)$$

In an equation, you can write either expression first. Thus, the following equation also represents the pattern.

$$a(b + c) = ab + ac$$

In this form, the equation is called the **distributive property.** There are many different forms of the distributive property. Two of the most important forms are given below.

> **The Distributive Property**
> For all numbers a, b, and c,
>
> $$a(b + c) = ab + ac$$
>
> and
>
> $$a(b - c) = ab - ac$$

Certain equations in algebra are called **basic properties.** A basic property is true for every value you substitute. Of course, it is impossible to try all the values to make sure. However, if the equation is false for even a single substituted value, then it is not a basic property. To show that $a - b = b - a$ is not a basic property, substitute 2 for a and 1 for b. This substitution is called a **counterexample.**

Critical Thinking   Which of the following statements are basic properties? Explain your answer.

a. $a + b = b + a$ **b.** $(x - y) - z = x - (y - z)$

Formulas

A **formula** is a general rule or principle representing a real situation or application. A formula is written in the form of an equation. For example, the formula stating the rule that the diameter of a circle is twice its radius is written as follows:

$$d = 2r$$

A formula is just *one* possible way to state a rule. The following diagram gives three additional ways to give the same rule.

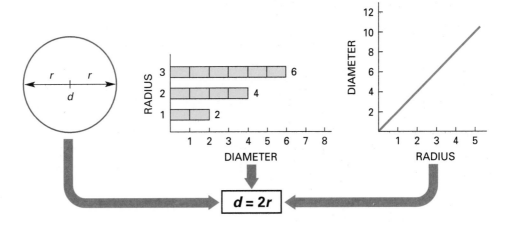

A formula such as $d = 2r$ is often the shortest, easiest way to write the relationship between two or more quantities. You have already worked with formulas such as these:

$$V = e^3 \qquad p = 2l + 2w$$

Formulas are shortcut ways to write mathematical rules or relationships. Where do formulas come from? For each formula, some person discovered the relationship that is expressed by the formula.

For example, people have known most of the formulas for finding areas since ancient times. Here are three formulas involving the area (A), the length (l), and the width (w) of a rectangle.

Find the area if you know the length and width: $A = lw$

Find the length if you know the area and width: $l = A \div w$

Find the width if you know the area and length: $w = A \div l$

1 Use the area formula to find the missing dimensions.

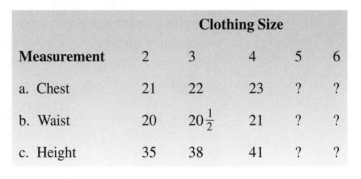

7

$A = ?$ 5

Given $l = 7$,
and $w = 5$, find A.

$l = ?$

$A = 35$ 5

Given $A = 35$,
and $w = 5$, find l.

7

$A = 35$ $w = ?$

Given $A = 35$,
and $l = 7$, find w.

2 Write three different problems related to a construction project that use the area formula in each of its forms.

Ongoing Assessment

The formula for the area of a rectangle can be used to find the area of a square. A square is a rectangle with equal sides. Thus, the formula for finding the area of a square with side, s, is

$$A = s^2$$

What is the area of a square with sides of 4 centimeters?

ACTIVITY 4 **Writing a Formula**

1 Complete the following chart. The measurements are in inches.

	Clothing Size				
Measurement	2	3	4	5	6
a. Chest	21	22	23	?	?
b. Waist	20	$20\frac{1}{2}$	21	?	?
c. Height	35	38	41	?	?

2 Write formulas to describe each pattern in the chart.

How do you know when to use a formula? And more importantly, how do you know which formula to use?

You can write the formula converting temperatures between the Celsius (C) and Fahrenheit (F) scales in two ways.

$$F = \frac{9}{5} C + 32 \qquad C = \frac{5}{9}(F - 32)$$

A temperature of 36°F converts to what equivalent temperature on the Celsius scale?

First, choose the version of the formula that has the variable you need on the left side of the equal sign. Since you are looking for a temperature in °C, choose the second version of the formula.

$$C = \frac{5}{9}(F - 32)$$

Substitute the numerical value in the problem for the variable it matches in the formula. Since the Fahrenheit temperature is 36°,

$$C = \frac{5}{9}(36° - 32°)$$

Do you recognize the procedure here? In an earlier lesson, Tracy used this formula to find the temperature in a computer room.

Use your calculator to evaluate the formula. If you round your answer to two decimal places, you should get C = 2.22. How did you do? Did you use the parentheses keys? Can you find the value for C without using the parentheses keys? Do not forget the units in the answer. 36°F is about 2.22°C.

WORKPLACE COMMUNICATION

Modern Workplace has 915 employees. Which plan should Mr. Morgan choose for the first month's paper supply? What else must he consider?

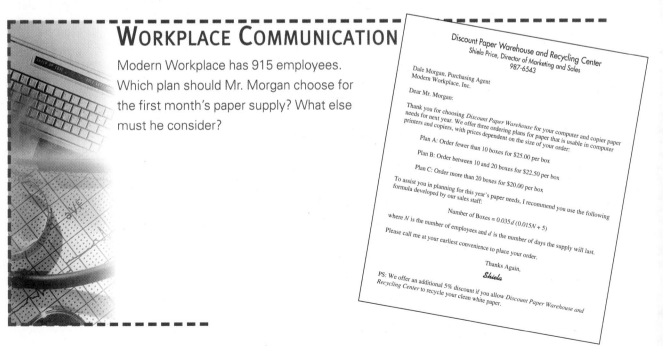

Discount Paper Warehouse and Recycling Center
Shiela Price, Director of Marketing and Sales
987-6543

Dale Morgan, Purchasing Agent
Modern Workplace, Inc.

Dear Mr. Morgan:

Thank you for choosing *Discount Paper Warehouse* for your computer and copier paper needs for next year. We offer three ordering plans for paper that is usable in computer printers and copiers, with prices dependent on the size of your order:

Plan A: Order fewer than 10 boxes for $25.00 per box

Plan B: Order between 10 and 20 boxes for $22.50 per box

Plan C: Order more than 20 boxes for $20.00 per box

To assist you in planning for this year's paper needs, I recommend you use the following formula developed by our sales staff:

Number of Boxes = $0.035d\,(0.015N + 5)$

where N is the number of employees and d is the number of days the supply will last.

Please call me at your earliest convenience to place your order.

Thanks Again,

Shiela

PS: We offer an additional 5% discount if you allow *Discount Paper Warehouse and Recycling Center* to recycle your clean white paper.

LESSON ASSESSMENT

Think and Discuss

1 What are the similarities and differences between an expression and an equation?

2 How can you tell whether or not an equation is a basic property?

3 How does a formula differ from an equation or a basic property?

4 Explain how to find the area of a square if you know the length of one side.

Practice and Problem Solving

Tell whether or not each equation is a basic property.

5. $a + (b + c) = (a + b) + c$

6. $a + 0 = 0$

7. $ab = ba$

8. $a \div b = b \div a$

9. $(a + b)^2 = a^2 + b^2$

10. $(a + b)^2 = a^2 + 2ab + b^2$

11. $-(a + b) = -a + (-b)$

12. $a - (b - c) = (a - b) - c$

13. $a(b - c) = ab - ac$

14. $a - b = a + (-b)$

Use a formula to find the area, length, or width of a rectangle with the given dimensions.

	Area	Length	Width
15.	?	48 mm	8 mm
16.	36 m^2	9 m	?
17.	108 mi^2	?	9 mi

18. Wallpaper comes in rolls that are 50 feet long and 18 inches wide. How many rolls of wallpaper will it take to cover 450 square feet of wall space?

19. A living room is 15 feet by 15 feet. A rug 12 feet by 12 feet is placed in the center of the room. How much of the floor is not covered by the rug?

20. The Deck Company is bidding on staining a rectangular deck that is 18 feet by 30 feet. If it bids on the job at $5 per square yard, what is The Deck Company's bid?

21. The weather report for the 4th of July is 35 degrees Celsius. What is the Fahrenheit temperature?

22. Water boils at 212 degrees Fahrenheit. What is the boiling point of water in degrees Celsius?

23. Which is hotter: 104°F or 40°C?

24. How can you convince someone that 0°C is the same as 32°F?

If D represents the distance you travel, R represents the rate you are moving, and T represents the time you travel, write a formula for

25. Distance.   **26.** Rate.   **27.** Time.

Use the formulas you wrote to solve these problems.
28. If it takes 4 hours to travel 240 miles, how fast are you driving?

29. If you run 3 hours at 5 miles per hour, how far will you run?

30. If a plane is flying at 240 miles per hour, how long will it take to fly 720 miles?

Mixed Review

Richard is using this population chart for statistical records.

Austin	465,632
Charlotte	395,934
Cleveland	505,616
Long Beach	429,321
New Orleans	496,938
Seattle	516,259

31. Rank the cities in order from least population to greatest population.

32. Use integers to write the difference between the population of each city and a benchmark population of 500,000.

33. The diameter of a molecule of water is approximately 0.0000000276 centimeters. Write the diameter in scientific notation.

34. Compute your age in seconds as of today at noon. Write your age in scientific notation.

35. The thickness of a page in a novel is about 8×10^{-5} meters. How thick is a 1200-page novel?

LESSON 3.3 CIRCLES

You can see circles in many of the designs found in architecture and in nature.

Circumference

The distance across a circle through the center is the **diameter** of the circle. The **circumference** is the distance around the circle. Complete the Activity to see what all circles have in common. You will need some string and a ruler for this Activity.

ACTIVITY | **The Ratio of Diameter to Circumference**

1 Find several circular objects.

2 Use your string and a ruler to find the circumference and the diameter of each object.

3 For each circle, divide the circumference by the diameter. Round the answer to the nearest hundredth.

4 Check with your classmates. What constant number can be used as the approximate quotient of your divisions?

5 This constant number is called π. To the nearest hundredth, π is equal to 3.14.

The Activity leads directly to the formula for finding the circumference (C) of a circle when you know its diameter (d).

Circumference of a Circle
  $C = \pi d$

Your calculator should have a π key. Press it to find what value it displays. Suppose a pipe has a diameter of 4.5 centimeters. To find the circumference of the pipe, follow these steps.

Enter 4.5

Press $\boxed{\times}$

Press $\boxed{\pi}$

Press $\boxed{=}$

Round the product to 14.14. The circumference of the pipe is about 14.14 centimeters. The symbol $\approx$ is used when the answer is not exact. Thus, you can write

$$C \approx 14.14 \text{ cm}$$

This is read "C is approximately equal to 14.14 centimeters."

Ongoing Assessment

You are insulating a replacement section of air conditioning ductwork. The ductwork is circular and has a diameter of 14 inches. What length of insulation is needed to wrap around the duct?

Critical Thinking   Remember that the diameter is twice the length of the radius. What is the formula for the circumference of a circle with radius r?

Area

The **area** of a circle is a measure of the space inside the circle. The area (A) of the circle is determined by its radius (r).

Area of a Circle
$$A = \pi r^2$$

Area is measured in square units such as square feet (ft^2) and square meters (m^2). The area of a circle with a radius of 1 meter is 3.14 square meters. The circumference of the same circle is 6.28 meters.

A circular swimming pool is 32 feet in diameter. A 5-foot wide sidewalk around the pool requires a nonskid surface. What area must the nonskid surface cover?

CULTURAL CONNECTION

Many ancient civilizations discovered how to find the area of a circle. Some Oriental cultures used 3 for the value of π. The Egyptians recorded a value close to 3.1604 in the Rhind papyrus written about 1600 BCE.

One of the closest approximations was used by the Greek mathematician, Archimedes, about 287 BCE. He determined that π must be

between $3\frac{1}{7}$ and $3\frac{10}{71}$. The Egyptian scientist Ptolemy used 3.1416 in about 150 AD. But it was not until the 18th century that it was finally proven that π cannot be written as a fraction or rational number. For this reason, π is called an **irrational number.** Today, computers have found π to enough places to fill a book. Find the value of π to nine decimal places.

LESSON ASSESSMENT

Think and Discuss

1 Explain the meaning of radius, diameter, and circumference of a circle. Draw a circle and label each term.

2 Explain how to find the circumference of a circle if you know the radius.

3 Explain how to find the area of a circle if you know the radius.

4 Which fraction, $\frac{22}{7}$ or $\frac{223}{71}$, is a better approximation for π?

Practice and Problem Solving

Find the circumference and area to the nearest hundredth of each circle having the given radius.

5. 2 meters　　**6.** 4 feet　　**7.** 8 centimeters

8. 2.7 yards　　**9.** 6.3 millimeters　　**10.** 1.25 kilometers

Find the area of each shaded region.

11. 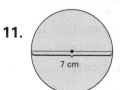 7 cm

12. 2 ft 6 ft

13. 4 m

14. The diameter of a truck wheel is 120 centimeters. What is the circumference of the wheel?

15. A patio is designed in the shape of a semicircle. If the radius of the patio is 10 feet, what is the area of the patio?

16. A pipe has a diameter of 3 inches. A welder must connect a wire clamp around the outside of the pipe. Find the length of the wire needed to wrap around the pipe.

Mixed Review

The average depth of the Pacific Ocean is 12,925 feet below sea level. The average depth of the Atlantic Ocean is 11,730 feet below sea level.

17. Write each depth as an integer.

18. What is the difference in the average depths of the two oceans?

19. The diameter of some white blood cells is about 4.03×10^{-4} inches. Suppose 1 million of these cells are aligned. Write the length of the line in scientific notation.

20. The force (F) on an object is found from the product of its mass (m) and acceleration (a), or $F = ma$. Find the mass of an object, in kilograms, when a force of 29.4 Newtons accelerates the object at a rate of 3 meters per second per second.

Boxes

The dimensions of a rectangular box are length (l), width (w), and height (h). The **volume** (V) of a box is the amount of space contained in the box. The volume is also the capacity of the box. In other words, the volume or capacity tells you how much the box will hold. If length, width, and height are equal, the box is a **cube.**

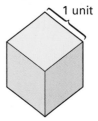

1 unit

The basic measurement for volume is a cube that measures 1 unit on an edge. This cube is 1 cubic unit. If the measure of the edge is in centimeters, the cube has a volume of 1 cubic centimeter (cm³).

1 cubic unit

Volume is measured by the number of cubic units an object can hold. Here is how you find the number of cubic centimeters in a box.

Find the number of cubic centimeters in the base.

Multiply by the number of layers in the box.

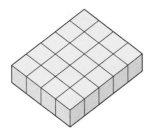

$5 \cdot 4 = 20$

$5 \cdot 4 \cdot 3 = 60$

The base contains 20 cubes, each 1 cubic unit.

The volume is 60 cubic units.

The picture leads to a formula for finding the volume of a box.

Volume of a Rectangular Box
$$V = lwh$$

If the box is a cube ($l = w = h = e$),

$$V = e^3$$

The inside of a rectangular container is 50 inches in length and 20 inches in width. If you fill the container to a depth of 8 inches, what is the volume of the container?

Critical Thinking   The meter is considered a one-dimensional measurement. Why is the square meter considered a two-dimensional measurement? Why is the cubic meter considered a three-dimensional measurement?

Spheres

What do basketballs, globes, ball bearings, storage tanks, and weather balloons have in common? All of these objects have the familiar shape of a sphere. The volume (V) of a sphere depends on the length of its radius (r).

Volume of a Sphere
$$V = \frac{4}{3}\pi r^3$$

A technician for the National Weather Service is preparing to launch weather balloons. Each balloon is spherical in shape and has a radius of 3 feet. To calculate the number of cubic feet of helium in each inflated balloon, the technician uses the volume formula.

$$V = \frac{4}{3}\pi r^3$$
$$= \frac{4}{3}\pi(3)^3$$
$$= \frac{4}{3}\pi(27)$$
$$= 36\pi \approx 36(3.14) \approx 113.1$$

Each balloon uses about 113.1 cubic feet of helium.

Find the volume of a spherical tank with a radius of 9 meters.

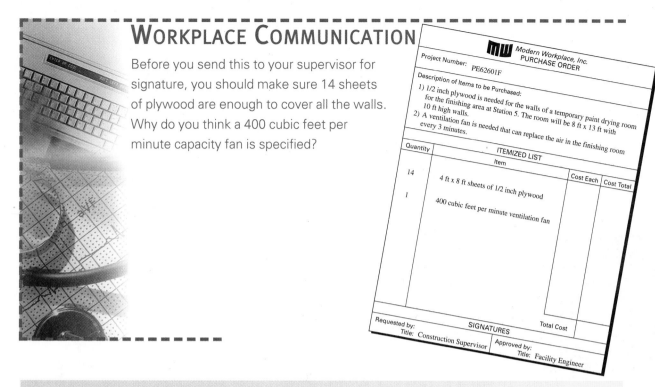

WORKPLACE COMMUNICATION

Before you send this to your supervisor for signature, you should make sure 14 sheets of plywood are enough to cover all the walls. Why do you think a 400 cubic feet per minute capacity fan is specified?

Modern Workplace, Inc.
PURCHASE ORDER

Project Number: PE62601F

Description of Items to be Purchased:
1) 1/2 inch plywood is needed for the walls of a temporary paint drying room for the finishing area at Station 5. The room will be 8 ft x 13 ft with 10 ft high walls.
2) A ventilation fan is needed that can replace the air in the finishing room every 3 minutes.

Quantity	ITEMIZED LIST Item	Cost Each	Cost Total
14	4 ft x 8 ft sheets of 1/2 inch plywood		
1	400 cubic feet per minute ventilation fan		

			Total Cost
Requested by: Title: Construction Supervisor	SIGNATURES	Approved by: Title: Facility Engineer	

LESSON ASSESSMENT

Think and Discuss

1 How are area and volume alike? How are they different?

2 How do you find the volume of a cube if you know the length of one edge?

3 How do you find the volume of a sphere if you know the diameter of the sphere?

Practice and Problem Solving

4. Find the volume of a box with length 3 meters, width 6.2 meters, and height 1 meter.

5. Find the volume of a cube with one edge 2.5 feet in length.

6. Find the volume of a sphere with a radius of 2.1 feet.

7. Find the volume of a sphere with a diameter of 6 meters.

8. Concrete costs $35 per cubic yard. If a concrete patio is to be 36 feet long, 12 feet wide, and 6 inches deep, how much will the concrete cost?

9. The diameter of the earth is about 7926 miles. Find the volume of the earth. Write your answer in scientific notation.

10. A steel ball has a diameter of 2 inches. It is placed inside a hollow sphere with a radius of 5 inches. How much water can be poured into the sphere?

11. A scoop of ice cream is placed on a cone. If the radius of the ice cream is 4 centimeters, what is the volume of ice cream?

30 in.

20 in.

12. A rectangular piece of cardboard is 30 inches by 20 inches. A piece of cardboard 2 inches by 2 inches is cut from each corner. The resulting piece of cardboard is folded to make a rectangular container. Find the volume of the container.

13. How many cubic inches equal 2 cubic feet?

14. How many cubic centimeters equal one cubic meter?

Mixed Review

You are planning a vacation. You want to rent a car for four days and drive 250 miles. The On-Time Rental company charges a one-time cost of $45 plus 35 cents a mile. The Best Deal company charges a flat rate of $37 per day.

15. Which company should you use?

16. How much will you save by using this company?

A business shows a loss of $295 in January, a gain of $1655 in February, and a loss of $695 in March.

17. Write each amount as an integer.

18. In which month did the business have the greatest loss?

19. What is the total gain or loss for the three months?

20. The Mississippi River has a drainage area of about 1,150,000 square miles. Write this drainage area in scientific notation.

LESSON 3.5 INTEREST

During the next few years, you will either save money or borrow money. When you keep money in a savings account, the bank pays you *interest* for the use of the money. When you borrow money, you pay *interest* for using money. **Interest** is money paid for the use of money.

The money that is borrowed or saved is called the **principal.** The **rate of interest** is the percentage of the principal charged for the use of the money. The rate of interest is usually specified for one year. However, you can calculate interest for any period of time.

Simple Interest

Simple interest is only earned or charged on the principal. Simple interest (i) is found by multiplying the principal (p) by the annual rate of interest (r) and the period of time (t), in years, that the money is borrowed or saved.

> **Simple Interest**
>
> $$i = prt$$

Suppose you borrow $500 for one year at 12% simple interest. To find the interest charged, use the formula $i = prt$ and multiply $500 \cdot 12\% \cdot 1$.

Remember to change the percentage to a fraction or a decimal before you multiply. That is, 12% is $\frac{12}{100}$ or 0.12. Find the product on your calculator. The simple interest is $60.

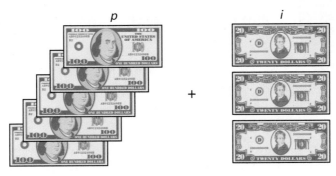

p i

$+$ $=$ **$560.00**

When you borrow money, you must pay back the **total amount** (*A*) of the loan. The amount is the principal plus the interest. The amount of the payback is $500 + $60 or $560.

Critical Thinking How is simple interest calculated if the money is borrowed for less than one year? More than one year?

Ongoing Assessment

If only one payment is made at the end of 3 years, what is the total payback on a loan of $750 borrowed at an annual rate of 9.5%? ▪▪▪

Compound Interest

Suppose you save $500 for two years at 8% simple interest. After one year, you will have $540 in the bank. After the second year, you will get another $40 interest and have $580. If you save the money at a bank, however, you earn more than simple interest. **Compound interest** is interest paid on both the principal and the previously paid interest. Using compound interest (compounded yearly), the amount in the bank after one year is still $540. For the second year, however, the $540 becomes the principal. The interest for the second year is calculated using this new principal.

$$i = prt$$

$$540 \bullet 8\% \bullet 1 = 43.20$$

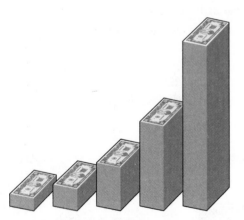

Instead of $580 after the second year, you now have $583.20. This is not a large increase over simple interest. However, over many years this amount will grow substantially. You can earn a great deal of money if you are paid compound interest instead of simple interest.

Total Amount with Compound Interest
$$A = p(1 + r)^n$$

A is the total amount and *n* represents the number of years.

Most banks compound interest on a daily rather than on a yearly basis. However, if interest is compounded even monthly, you receive much more money over a period of years than simple interest provides.

Using a Power Key

A bank is advertising a plan that encourages parents to invest money at 8% interest compounded yearly when their child enters first grade. If a parent invests $500 under this plan, how much money will the child have 12 years later?

The formula for compound interest is the following:

$$A = 500(1 + 0.08)^{12}$$

1 Use the power key on your calculator.

Enter 500	Press $\boxed{\times}$
Enter 1.08	Press $\boxed{y^x}$
Enter 12	Press $\boxed{=}$

2 What is the result?

3 How much less is the same amount invested at simple interest?

LESSON ASSESSMENT

Think and Discuss

1 Compare the amounts you receive when you save $100 at 6% simple interest and $100 at 6% compounded annually for 10 years. Why is saving at compound interest important?

2 A credit card company charges 18% annual interest on any unpaid balance. Explain why this is not such a good way to borrow money.

Practice and Problem Solving

Find the difference for each amount of savings if simple interest or compound interest is used.

3. $1000 at 6% for 2 years.   **4.** $875 at 5% for 5 years.

5. $2,450 at 9% for 3.5 years.   **6.** $750 at 8.5% for 4 years.

7. Todd invested $750 at 6.5% compounded annually. How much did Todd have after 5 years.

8. Kesha borrowed $1800 at 5% simple interest. How much did Kesha owe after 6 months?

9. Angie put $2000 in a CD. The CD paid 6% compounded annually. How much is the CD worth after 3 years?

10. Suppose you invest $1000. How long will it take to double your money at 6% simple interest?

Mixed Review

A clothing company sells four lines of clothing. The chart shows the profit (P) and loss (L) in dollars for each line over one year.

Line A	Line B	Line C	Line D
220,000(P)	105,000(L)	92,500(L)	162,550(P)

11. Which line has the greatest loss?

12. Which line has the least profit?

13. Draw a number line and place each profit and loss in the correct position.

Write the product or quotient in scientific notation.
14. $(3.4 \times 10^3)(6.2 \times 10^{-5})$

15. $(1.25 \times 10^{-2}) \div (3 \times 10^{-3})$

16. What is the value of $m - (n - m)^2$, when m is 6 and n is 2?

17. A square picture is 3 feet on each side. It costs $15 to matte each square foot. How much does it cost to matte the picture?

18. A washer has an inside radius of 15 millimeters and an outside radius of 21 millimeters. What is the area of the washer?

19. A grain car is 30 feet long, 10 feet wide, and 8 feet high. One cubic foot of grain weighs 6 pounds. What is the weight of the grain needed to fill the car?

20. Find the volume of a sphere with a radius of 12 feet.

Math Lab

Equipment Calculator
Timer
Masking tape
Battery-powered cars
Tape measure

Problem Statement

You will measure the time required for battery-powered cars to travel a measured distance. You will then calculate the speed of the cars. Based on the comparison, you will predict the winner of a race. An actual race will test the validity of your prediction.

Procedure

a Place a 1-foot piece of tape on the floor as a starting line. Measure 15 feet and place another 1-foot piece of tape for a finish line.

b Form a team to work with each car. Each team should measure the time required for its car to go from the starting line to the finish line. Write this time on a sheet of data paper. Repeat and record the measurements for a total of five trial runs for each car.

c Use an appropriate formula to calculate the speed of the car for each trial run in feet per second (ft/sec). Round each answer to the nearest hundredth of a foot per second.

d Calculate the average speed of the car for the five trial runs to the nearest hundredth of a ft/sec.

e When all groups have finished recording results, compare the average speeds and predict which team's car will win a race.

f Measure a distance of 25 feet. Race the cars over the 25-foot distance. Record which cars take the first three places.

g How do the results of the race compare with the predictions made in Step **e**?

Activity 2: Radius and Volume of a Sphere

Equipment Calculator
Micrometer caliper
100 BBs
Five $\frac{3}{8}$ inch ball bearings
10 ml graduated cylinder

Problem Statement

You will measure the diameter of two different spheres and calculate their volumes. Then, you will measure the volumes of a number of the spheres by the water-displacement method. From this data, you will calculate an average volume for each sphere and compare the average volume to the calculated volume. You will also calculate the radius of a sphere given its volume.

Procedure

a Use the micrometer to measure the diameter of a $\frac{3}{8}$ inch ball bearing. Convert this measurement from inches to centimeters. Record this measurement to the nearest 0.001 cm.

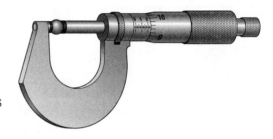

b Calculate the volume of the ball bearing using the formula for the volume of a sphere. Remember, the radius is one-half of the diameter.

c Fill the 10 ml graduated cylinder with water to about the 5 ml mark. Read the initial volume of the water. Record this initial volume.

d Place the ball bearings in the 10 ml graduated cylinder one at a time until all five are added. If any of the ball bearings are above the final water level, start over with more water. Read and record the final volume of the water.

e Subtract the initial volume from the final volume. Since a millimeter is the same as a cubic centimeter, write the volume change in both units. Divide this volume change by the number of ball bearings added. This is the average volume of one ball bearing.

f Compare the average volume measured in Step **e** to the calculated volume in Step **b**.

g Use the average volume calculated in Step **e** and the formula below to find the average radius of each ball bearing.

$$r = \sqrt[3]{\frac{3V}{4\pi}}$$

The symbol $\sqrt[3]{}$ means "cube root." For example, $\sqrt[3]{8}$ is the number that answers the question, "What are the three equal numbers whose product is 8?" In other words, $\sqrt[3]{8} = 2$ because $2 \times 2 \times 2 = 8$.

Most scientific or graphics calculators can be used to find a cube root.

See if either of these methods will find the cube root of 216.

1. If your calculator has a $\boxed{\sqrt[3]{}}$ key, enter 216 and press $\boxed{\sqrt[3]{}}$.

2. If your calculator has a $\boxed{y^x}$ key, an $\boxed{\text{INV}}$ key, a $\boxed{\sqrt[x]{y}}$ or a key labeled $\boxed{y^{1/x}}$, enter 216 and press the necessary keys.

Try either method with 256. You should get 6.35 rounded to the nearest hundredth. Now continue with the Math Lab.

Compare this average radius to $\frac{1}{2}$ the diameter of the ball bearing measured with the micrometer calipers in Step **a**.

h Use the micrometer to measure the diameter of a BB. Convert this measurement from inches to centimeters. Record this measurement to the nearest 0.001 cm.

i Calculate the volume of the BB using the formula for the volume of a sphere.

j Fill the 10 ml graduated cylinder with water to about the 5 ml mark. Read and record the initial volume of the water.

k Add the BBs to the graduated cylinder, one at a time, until all 100 are added and all are below the final water level. Read and record the final volume of the water.

l Subtract the initial volume from the final volume. Divide this volume change by the number of BBs added. This is the average volume of a BB.

m Compare this average volume of a BB to the calculated volume from Step **h.**

n Use the average volume calculated in Step **l** and the formula for the radius of a sphere to calculate an average radius. Compare this average radius to $\frac{1}{2}$ the diameter of the BB you measured with the micrometer in Step **h.**

o How do the average measured volumes compare to the calculated volumes? How do the average radii compare to $\frac{1}{2}$ the measured diameters?

p Which method do you think is more accurate?

1. The one based on measurement with a micrometer.

2. The one based on water displacement.

Activity 3: Indirect Measurement of Height

Equipment   Calculator
Protractor
PVC pipe-1 inch in diameter and 3 feet long
   with a swivel attached in the center.
Line level
Tape measure

Over the length of the pipe, draw a line through the swivel. Attach 20 feet of string to the swivel at point **A.**

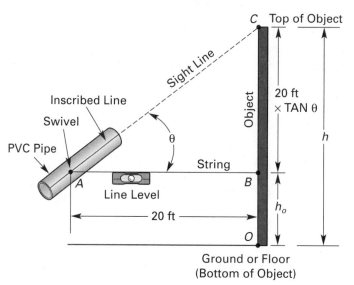

Problem Statement

You will use the device in the figure and a formula to determine the vertical height of an object perpendicular to the ground.

Procedure

a Form a five-member group and examine the figure.

- Member One should sight through the pipe along the line *AC*.

- Member Two should be ready to measure angle θ.

- Member Three needs to check the line level along *AB* to ensure the string *AB* is level.

- Member Four needs to hold the end of the string against the measured object at point *B*.

- Member Five needs to measure the height h_0 of the string *AB* above the base (point *O*) of the object.

b Study the drawing and examine the following formula:

$$h = 20(\tan \theta) + h_0$$

The expression tan θ is read "tangent of theta." You will learn more about the tangent in a later chapter. For now, you just need to know that the tangent is the ratio of the side opposite angle θ to the side adjacent to angle θ in a right triangle. In the drawing, tan θ is the ratio of the length of the segment $(h - h_0)$ to the length of the horizontal string.

Use a scientific calculator to find the value of *h* using the formula. Enter the measure of the angle θ in degrees and press the (TAN) key. The value for tan θ will appear in the display.

Suppose θ is 45° and h_0 is 4 feet.

Enter 45° and press the (TAN) key. The display should read 1.0000000. If the display does not read 1.0000000, change your calculator to the degree mode. When you get a value of 1.0000000 for tan 45°, multiply it by 20 and add 4 to this result. This gives a value of 24 feet for the height *h*.

c Select an object to measure. Some good choices outside your classroom are a flagpole, a wall of the school building, or a football goalpost. Inside the classroom, you can measure the height of a wall.

d Hold the free end of the 20-ft string against the selected object at point B. Stretch the string tight and attach the line level. Measure the length AB from the swivel to point B to verify that it is 20 ft. If AB is not 20 ft, use the true measurement AB instead of 20 in the formula for h.

e Hold the pipe at a comfortable height. Sight through the pipe to the top of the object.

f Adjust the end of the string at point B until the line level shows that the string is level. Measure and record the height OB. In the formula, this measure is called h_0.

g While sighting through the pipe to the top of the object, measure and record the angle of elevation, θ, between the string and the line inscribed on the pipe at the swivel.

h Use the appropriate formula to calculate the height of the object, h. Record this value.

i Describe other methods you could use to determine the height of the object. Which method do you think is most accurate?

j Describe how you would change the procedure to measure the width of a wide river or canyon.

MATH APPLICATIONS

The applications that follow are like the ones you will encounter in many workplaces. Use the mathematics you have learned in this chapter to solve the problems. Wherever possible, use your calculator to solve the problems that require numerical answers. A table of conversion factors can be found beginning on page A1.

An automotive technician needs to determine the resistance of an automobile starting motor that draws 90 amperes of current from a 12 volt battery.

1 Use the formula $R = \frac{V}{i}$ to calculate resistance, where R is the resistance (ohms), V is the voltage across the resistance (volts), and i is the current (amperes).

A cleaning solvent is purchased by the barrel. A useful formula for estimating the number of barrels in a cylindrical storage tank is

$$V = 0.14d^2\,h$$

V = volume of solvent in barrels.
d = diameter of the storage tank in feet.
h = height of the storage tank in feet.

2 Approximately how many barrels of cleaning solvent can be stored in a 20-foot diameter tank that is 12 feet tall?

3 You are preparing a purchase order to refill the storage tank with solvent. The depth of the solvent remaining in the tank is 7 feet. How many barrels of solvent should you order?

Bricklayers can estimate the number of standard bricks needed to build a wall using the formula $N = 21\,LWH$.

N = number of bricks needed.
L = length of the wall in feet.
W = width of the wall in feet.
H = height of the wall in feet.

4 How many bricks are needed to construct a wall that is 150 feet long, 12 feet high, and 4 inches wide?

A construction team is pouring concrete for six cylindrical bridge supports. Each of the supports has a 2.4 foot radius and is placed into the ground to a depth of 12 feet and extends 15 feet above the ground.

5 What is the total length of each support in yards?

6 Use the formula

$$V = \frac{\pi d^2 L}{4}$$

where V is the volume of the cylinder, d is the diameter of the cylinder, and L is the length of the cylinder in yards to find the number of cubic yards of concrete needed to pour one of the supports.

7 A truck can deliver eight cubic yards of concrete with each load. How many trucks of concrete will the contractor need to pour the six supports?

Suppose that your time card shows that you worked seven hours each day, for five days. Your rate of pay is $7.45 per hour. You can determine your gross pay for a week by multiplying the total hours worked during the week by your hourly rate of pay.

8 Write a formula that determines weekly gross pay.

9 Use the formula to compute your gross pay.

10 If you work more than 40 hours during a week, you are paid 1.5 times the normal hourly pay rate for those hours worked in excess of 40 hours. Write a formula to determine your gross pay earnings for work in excess of 40 hours.

Your medical insurance policy requires you to pay the first $100 of your hospital expenses (this is known as a deductible). The insurance company will then pay 80% of the remaining expenses. The formula below expresses the amount that you must pay.

$$E = [(T - D) \times (1.00 - P)] + D$$

E is the expense to you (how much you must pay).
T is the total of the hospitalization bill.
D is the deductible you must pay first.
P is the decimal percentage that the insurance company pays after you meet the deductible.

11 Suppose you are expecting a short surgical stay in the hospital for which you estimate the total bill to be about $5000. Use the formula above to estimate the expense to you for the stay in the hospital.

Aerial photography is a useful tool in forestry surveys. A photograph can be used to estimate the diameter of a ponderosa pine. The diameter at chest height is found using the following formula:

$$D = 3.7600 + (1.3480 \times 10^{-2})\ HV - (2.4459 \times 10^{-6})\ HV^2 + (2.4382 \times 10^{-10})\ HV^3$$

D is the diameter at chest height in inches.
H is the height of the tree in feet.
V is the visible crown diameter from the photograph in feet.

12 An aerial photograph shows a visible crown diameter of 22 feet for 108-foot trees. Find the diameter of the trees.

When determining the usable log volume, deductions are made for slab, edgings, and saw kerf. The following formula is used to estimate the number of board feet that can be obtained from a log.

$$V = 0.0655\ L\ (1 - A)\ (D - S)^2$$

V is the usable volume of a log in board feet.
L is the length of the log in feet.
A is the decimal percent deduction for saw kerf.
D is the log diameter in inches.
S is the slab and edging deduction in inches.

13 Use the formula to estimate the usable volume of a 32-foot log having a diameter of 30 inches. Allow for 10% saw kerf and 3 inches of slab and edging deductions.

A commercial cherry grower estimates that when up to 30 trees are planted per acre, each tree can produce about 50 pounds of cherries per season. For each additional tree planted, the yield per tree is reduced by 1 pound.

14 Make a table of the yield per tree for several different planting densities: 30 per acre, 31 per acre, 32 per acre, and so on, up to 35 per acre.

15 Add a column to your table that shows how many total pounds of cherries would be produced per acre for each planting density.

16 Someone proposes that these data can be described by the formula

$$Y = 50 - (D - 30)$$

Y is the yield per tree in pounds.
D is the density of the planting in trees per acre.

Does this formula agree with the data in your table?

A floating electrical stock-tank heater uses 1500 watts of electrical power. The electric company charges you 12¢ per kilowatt-hour of usage. This means that if you use a one-kilowatt device for one hour, the electric company will charge you 12¢ for the electricity. If you use it for two hours, you will be charged 24¢, and so on.

17 During the winter, you use the stock-tank heater at night. A timer will control the heater, so that it operates for about 14 hours each night. Write an equation to compute the cost of electricity to run this heater for d nights.

18 Use your equation to find the electricity cost of using the heater for 90 nights.

BUSINESS & MARKETING

To consider an investment in certificates of deposit, you use the formula

$$A = P(1 + r)^n$$

Note: The units of time for i, r, and n must agree (for example, if you use a yearly percentage rate for r, you must use years for n).

19 A long-term certificate requires a deposit of $5000 for five years. Each year, the deposit earns 8.5% interest. Use the formula to determine the future value of the deposit at the end of the five-year term.

20 A short-term certificate requires a deposit of $1000 for six months. The deposit earns interest at an annual rate of 7%, or a monthly rate of $\frac{1}{12}$ of 7%. Use the formula to determine the total amount of the deposit at the end of the six-month term.

Series *EE* Savings Bonds can be purchased for one-half their face value. If they are held until maturity, they can then be redeemed for their face value, plus a bonus. A securities teller must explain to a customer how to compute the interest obtained on these bonds. The difference between the redemption value and the purchase price is the interest earned.

21 Write a formula that shows how to compute the purchase price of a Series *EE* Savings Bond, knowing only its face value.

22 Write a formula that shows how to compute the interest earned on a Series *EE* Savings Bond held until maturity, knowing only the purchase price and the redemption value.

A delivery service limits the packages it will handle to those weighing less than 70 pounds. In addition, the combined girth and length of the package must not exceed 108 inches. The length is the longest dimension of the package, and the girth is the total distance around the package at its widest part, not including the length.

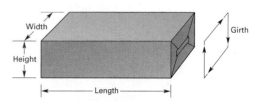

23 For a rectangular package with width *W*, length *L*, and height *H*, write a formula to compute the "combined girth and length."

24 You want to ship a small telescope in a box that weighs 54 pounds, and has a length of 54 inches, a width of 10 inches, and a height of 11 inches. Would this package pass the weight and size restrictions of the delivery service?

You can obtain a suggested retail price for used cars from a used-car guide. If the car has unusually low or unusually high mileage, the suggested price is adjusted accordingly. The following equation for cars driven between 20,000 and 70,000 miles seems to agree with your table of prices for a particular make and model of car.

$$P = -0.024\,M + 4850$$

P is the suggested retail price of the car.
M is the mileage shown on the car's odometer (in miles).

25 What is the suggested price for this type of car if the odometer shows 25,000 miles?

26 Does this equation suggest higher or lower prices for cars with unusually high mileage? What about very low mileage? Is this as you expect? What indicates this trend in the equation?

Some electric utilities offer an average billing plan to their customers. According to the plan, the amount a customer must pay on a given month's bill is determined as follows:

The current month's cost and the previous 11 months' costs are added, and this total is divided by 12. One customer's monthly utility costs are shown below.

Utility Costs for Customer #17662			
Month	**Cost**	**Month**	**Cost**
Current	$ 67.02	April	$76.15
September	92.66	March	70.28
August	126.90	February	55.23
July	127.43	January	58.18
June	112.76	December	49.72
May	89.19	November	59.14

27 Write a formula that allows you to compute this customer's bill using the average billing plan described above. The formula should require you to supply only the monthly costs to arrive at a bill for the current month. Use your formula to compute the bill for the current month.

You can make payments on an insurance policy either annually or monthly. If you choose to spread the premium payments over 12 months, the insurance company charges a handling fee of 11% of the annual premium. This handling fee is also spread over the 12 monthly payments.

28 Write a formula to determine the monthly payment for an annual premium plus the handling fee.

29 Use your formula to determine what the monthly payment would be if the annual premium is $288.

A client calls your investment firm to invest $20,000. She wants to earn at least $2400 interest per year by splitting the investment between two options. You can give her this choice: Option A pays 9% interest per year at relatively low risk, and Option B pays 17% per year at a higher risk. It is reasonable to keep the amount invested in the high-risk option as low as possible. The anticipated interest earned from each investment each year can be computed by multiplying the annual interest rate by the amount invested.

30 You advise your client to invest L dollars (L is less than $20,000) in the low-risk investment. Write an expression involving L that shows how much remains to invest in the high-risk investment.

31 Write a formula to compute how $2400 in interest can be earned from investing L dollars in the low-risk investment and the remaining amount in the high-risk investment.

32 How much of the $20,000 do you recommend the client deposit into each of the two investments?

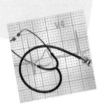

You are performing laboratory tests on blood samples to find the amount of oxygen in the hemoglobin of the red blood cells. A blood test tells you that a 100 milliliter sample contains 12 grams of hemoglobin, and the hemoglobin is 91% oxygen. The following formula is used to detemine the amount of oxygen in the blood.

$$H = 1.36gs$$

H is the milliliters of oxygen in the hemoglobin from 100 ml of whole blood.
g is the number of grams of hemoglobin in 100 ml of whole blood.
s is the decimal value of oxygen per cent in the hemoglobin.

33 How much oxygen is in the hemoglobin of the sample (round your answer to 2 decimal places)?

As an exercise consultant, you have a formula that helps you estimate the percentage of body fat for men.

$$F = 0.49W + 0.45P - 6.36R + 8.71$$

F is the percent of body fat.
W is the waist circumference in centimeters.
P is the thickness of the skin fold above the pectoral muscle, in millimeters.
R is the wrist diameter in centimeters.

34 Compute the percent body fat for a male client who has a waist circumference of 87.3 cm, a skin fold thickness of 6.2 mm, and a wrist diameter of 6.5 cm (round your answer to one decimal place).

35 Suppose your brother's waist measures 34 inches, the skin fold above his pectoral muscle is about one-half inch, and his wrist diameter is two and one-half inches. Estimate your brother's percent body fat.

In a hospital, the doctor frequently orders a volume V (in cc) of IV (intravenous) fluid to be given over a specified period of hours, T.

The nurse must convert the rate from "cc per hour" to "drops per minute" using a conversion ratio of "15 drops per cc of fluid." A formula tells the nurse how many drops per minute to set the IV:

$$D = \frac{V}{4T}$$

D is the number of drops per minute.
V is the volume of fluid (in cc) ordered by the doctor.
T is the total time (in hours) which the fluid is to be given.

36 Suppose the doctor orders 1000 cc of D_5W to be given to a patient over eight hours. How many drops per minute should the IV supply to the patient?

Blood-sample analysis in the laboratory requires you to compute the mean corpuscular volume, or MCV.

$$MCV = \frac{H}{RBC} \times 10^7$$

MCV is the mean corpuscular volume in cubic centimeters.
H is the hematocrit value expressed in percent (for example, if the red blood amount is 52%, use 52—not 0.52—in the formula).
RBC is the red blood cell count in cells per cc.

37 Evaluate the equation when H is 45% and RBC is 5.3×10^6 cells per cc.

FAMILY AND CONSUMER SCIENCE

You have $8750 in fire and smoke damage to your house. Your insurance policy provides for 80% copayment. Your property is valued at $65,000, and the insurance policy covers $50,000.

38 Use the formula

$$P = \frac{i}{0.80V} \cdot L$$

P is the payment made by the insurance company.
i is the insurance coverage of the policy.
L is the loss due to fire.
V is the value of the property that is insured.

to determine the payment you will receive from the insurance company.

A furniture company rents a sofa and love seat combination for an initial fee of $79.80 and a monthly rental payment of $64.80. The company's "rent-to-own plan," lets you keep the furniture after the 24th payment.

39 Write a formula for the total cost to you of the furniture rental after any payment before the 24th payment.

40 Evaluate your equation for the 24th payment. How much will you have paid for the furniture when it is yours to keep?

INDUSTRIAL TECHNOLOGY

Greenshield's formula can be used to determine the amount of time a traffic light at an intersection should remain green: $G = 2.1n + 3.7$. G is the "green time" in seconds, n is the average number of vehicles traveling in each lane per light cycle.

41 Find the green time for a traffic signal on a street that averages 19 vehicles in each lane per cycle.

The *"WHILE"* formula is a simplified method to estimate the power of an air conditioning unit needed for a given room. The formula is

$$BTU = W \ H \ I \ L \ E \div 60$$

where BTU is the number of *Btu* (British thermal units) per hour of cooling power needed for the room.

W is the width of the room in feet.
H is the height of the room in feet.
I is the insulation factor (10 for well insulated rooms and 18 for poorly insulated rooms).
L is the length of the room in feet.
E is the exposure factor.

You must determine the exposure factor from the orientation of the longest wall. If the wall faces north, $E = 16$; if east, $E = 17$; if south, $E = 18$; if west, $E = 20$. If two or more walls are equally the "longest," use the largest possible value for E.

42 A newly constructed room needs air conditioning. It is 18 feet wide, 24 feet long, and 9 feet high. The room is well insulated. One 24-foot wall faces south and one faces north. What size air conditioner does the formula predict is needed to cool this room (round your answer to the nearest 1000 *Btu* per hour)?

Skills

Evaluate each expression.

1. $9 + 16 \bullet 2$ **2.** $64 \div (16 \div 4)$ **3.** $8 - 3^2 + 9$

4. $5a - 7b$, when a is -3 and b is -4.

5. $m^3 + 6 - n^2$, when m is -2 and n is -3.

6. $pq^2 + r$, when p is 4, q is 3, and r is -10.

7. Give a numerical example of the distributive property.

Applications

8. The temperature on a bank thermometer shows 45°C. What is the temperature in degrees Fahrenheit?

9. A patient has a temperature of 100°F. What is the patient's temperature in degrees Celsius?

This table shows the number of plants growing per acre over five years.

Year	0	1	2	3	4	5
Plants	13.5	17.0	20.5	24.0	27.5	?

10. Write a formula to show the relationship between the number of plants growing per acre in a given year, and the number of plants growing per acre the following year.

11. Use the formula to complete the table for five years.

In Exercises 12–14, let A represent the area of a rectangle and P represent the perimeter. Let L represent the length and w represent the width of the rectangle.

12. Write the formula for finding the perimeter of the rectangle.

13. Write the formula for finding the area of the rectangle.

14. Write the formula for finding the length of the rectangle if you know its area and width.

15. What is the diameter of a circle that has a radius of 7.5 inches?

16. The path from the center of a circular garden to its edge is 8 feet long. What is the length of a path around the outer edge of the garden?

17. Suppose one quart of paint will cover 180 square feet. You need to paint a circular sign that has a diameter of 8 feet. How many quarts of paint must you purchase?

18. One cubic yard of concrete costs $48. A rectangular concrete pier has dimensions 4 feet by 2 feet. The pier is 18 feet high. How much will it cost to fill the pier with concrete?

19. The formula for the volume of a cylinder is $V = \pi r^2 h$, where r is the radius and h is the height. One cubic foot of water weighs about 62.4 pounds. What is the weight of the water in a cylindrical tank with a radius of 6 feet if the tank is filled to a depth of 12 feet?

20. What is the simple interest on a note for $2000 borrowed for 6 months at an annual rate of 8%?

Math Lab

21. Three battery-powered cars travel the following distances in the indicated times. Which car would win a three-way race?

Car	Distance (feet)	Time (seconds)
A	11	1.9
B	15	2.3
C	13	2.0

22. You measure the diameter of a sphere with a micrometer caliper. How would you use this measurement to find the volume of the sphere?

23. Find the height of a treetop if sighting through a pipe attached to a 20 foot string gives an angle (θ) of 60°. The string is 5 feet above the ground. $[h = 20(\tan \theta) + h_0]$?

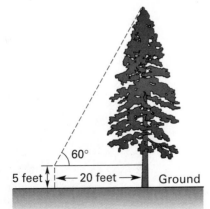

60°

5 feet   ←— 20 feet —→   Ground

CHAPTER 4

WHY SHOULD I LEARN THIS?

Equations are the backbone of today's technology. People use equations to control air pollution, design supersonic aircraft, and operate telecommunication systems. Learn how you will use equations in your future career.

SOLVING LINEAR EQUATIONS

OBJECTIVES

1. Simplify and solve an equation.
2. Translate a problem into an equation.
3. Check the solution of an equation in terms of the problem.

In Chapter 3, you used equations in the form of formulas. In this chapter, you will solve formulas and equations for a specific variable. Who uses equations? Most people who work with money use equations—people in banks, government tax offices, real estate firms, credit unions, and payroll departments. People who work in factories, in hospitals, on farms, and around the home also use equations. Here are two ways people use equations on the job.

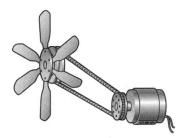

$B = 2L + 1.625 (D + d)$
designing an exhaust fan

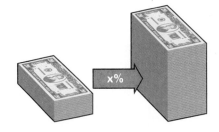

$A = P (1 + r)^n$
helping savings grow

How do equations help you solve problems? Often the problems you meet on the job are stated in words. If you can state a problem in the form of an equation, you can usually solve the equation and find an answer to the problem. Once you find an answer, it is important to check and analyze the solution in terms of the original situation or problem. This check will make sure your answer makes sense.

As you watch the video for this chapter, notice how people on the job use equations to solve problems.

LESSON 4.1 SOLVING MULTIPLICATION EQUATIONS

Understanding equations and how to rearrange them is an important skill in the workplace. Often a technician is asked to review technical reference books and find an equation for solving a particular problem.

Roland is an electronics technician. He knows the current and resistance specifications for a particular computer circuit but not the voltage. To find the voltage, he refers to a reference manual and finds the following equation:

$$V = I \cdot R$$

This equation is called Ohm's Law. In this equation, V represents the voltage, I represents the current, and R represents the electrical resistance. To find the voltage, Roland can multiply the measured values of the current by the resistance. The way this equation is arranged—with the V isolated on the left side—makes Roland's work very easy.

Suppose Roland is able to measure the voltage V and current I but needs to calculate the resistance R of the circuit. Can he still use Ohm's Law?

The answer is yes, but first Roland must rearrange the equation to "solve" for R.

Equations and Pan Balances

Remember, an equation states that two mathematical expressions are equal. For example,

$$V = IR$$

states that the expression V is equal to the expression IR.

One way to help you understand equations is to think about a pan balance. A pan balance is an instrument that measures the relative weights of objects.

Small weights are added to the right-hand pan until they balance the weight of the object in the left-hand pan. When "balanced," the weight of the object on the left equals the total weight on the right.

In the same way, an equation represents a balance between its two sides. The equal sign says the expression on the left side of the equation equals the expression on the right side. Consequently, even though one expression is represented by *V* and the other is represented by *IR*, they are equal, and both sides of the equation are balanced.

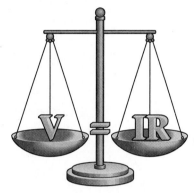

When solving equations, the most important thing to remember is that you must always keep the equation balanced. This balance is maintained by always performing the same mathematical operations on *both* sides of the equation.

> **Balancing Equations**
> Whatever mathematical operation is performed on one side of an equation must also be performed on the other side.

You can use this rule to solve multiplication equations.

Multiplication Equations

ACTIVITY | **Solving Multiplication Equations**

1 Start with the equation 5 • 3 = 15.

2 Divide each side by 5.

3 What is the result?

Now try this:

4 Start with the equation $5x = 15$.

5 Divide each side by 5.

6 What is the result?

You have just solved the multiplication equation $5x = 15$ by dividing each side of the equation by 5. The resulting equation is $x = 3$. Thus, the solution is 3. To solve the equation, you used one of the basic properties of equality.

> **Division Property of Equality**
> If each side of an equation is divided by the same nonzero number, the results are equal; that is, the two sides of the equation stay equal or balanced.

The Division Property solves problems modeled by multiplication equations. When solving equations using division, it is easier to show the division in fraction form rather than with the division sign. Thus, $5x \div 5$ is written as $\frac{5x}{5}$.

When you solve an equation, give your reason for each step. This will help you solve equations as they become more complicated.

EXAMPLE 1 Solving a Multiplication Problem

A shipping clerk must ship several packages of chemicals to a laboratory. The container holds 18 pounds. If a packet weighs 3 pounds, how many chemical packets will each container hold?

SOLUTION

The equation $3x = 18$ models this problem. Here is how to solve $3x = 18$.

$3x = 18$	Given
$\dfrac{3x}{3} = \dfrac{18}{3}$	Division Property
$x = 6$	Simplify

The last step is an equation with *x isolated* on one side. The solution, 6, is on the other side of the equal sign. Thus, each container can hold 6 chemical packets.

Ongoing Assessment

a. Solve the equation $4x = -24$. Give the reason for each step.

b. Solve the equation $6x = 21$. Give the reason for each step.

Critical Thinking   Why do you choose the coefficient of the variable as the divisor when you are solving a multiplication equation?

Now return to the example of Ohm's Law.

Roland has measured the voltage and current in a computer circuit and needs to find the circuit resistance. He knows Ohm's Law relates these circuit measurements to each other through the following equation:

$$V = IR$$

Roland uses the following steps to solve this equation for *R*. Give the reason for each step.

$$V = IR \qquad\qquad ?$$
$$\frac{V}{I} = \frac{IR}{I} \qquad\qquad ?$$
$$\frac{V}{I} = R \qquad\qquad ?$$

You can write the isolated variable on either side of the equal sign. However, the final result is usually written with the isolated variable on the left side.

$$\frac{V}{I} = R \text{ means the same as } R = \frac{V}{I}$$

In solving $V = IR$ for R, I is divided by itself. When any nonzero number is divided by itself, the result is 1.

$$\frac{IR}{I} = \frac{I}{I} \bullet R$$

$$R = 1 \bullet R$$

When you multiply R by 1, the result is R. This is an important basic property.

Property of Multiplying by 1
 For any number a,

$$a \bullet 1 = a$$

and

$$1 \bullet a = a$$

Ongoing Assessment

Solve the equation $C = \pi d$ for d. Show your steps. When did you use the Property of Multiplying by 1?

John is starting a new job. His employer has agreed to pay him $5.35 per hour. They also agree that John will work 15 hours per week after school and 8 hours on Saturdays.

To find his weekly pay, John uses the following equation:

weekly pay = $5.35/hr (15 hr + 8 hr)
                           ↓   ↓
                    hourly rate   total hours
                                            worked

Let p represent John's weekly pay. You can rewrite the last equation as

$$p = 5.35 (15 + 8)$$

The solution is 123.05. This means that John will earn $123.05 each week.

Now suppose John wants to increase his weekly earnings to $150 per week so that he can afford a stereo. But he still wants to work the same number of hours. The only

way John can increase his earnings without working more hours is to earn a higher hourly rate. What new hourly rate does John need?

To answer this question, first modify the last equation by replacing p with $150. Then let r represent the new hourly rate. The result is the following equation:

$$150 = r(15 + 8)$$

You can also write this equation in the following forms:

$$150 = r(23)$$

$$23r = 150$$

You can solve this equation by dividing both sides by 23. Why? The result is 6.52 rounded to the nearest hundredth. Thus, if John works 23 hours each week, he will have to earn $6.52 per hour to make $150 per week. Always check and analyze the solution to make sure it makes sense; that is, $6.52 \times 23 = \$149.96$.

The Division Property of Equality is also useful in solving geometry problems.

EXAMPLE 2 Finding the Degrees in a Triangle

A sheet metal worker is making a warning sign in the shape of an equilateral triangle. How many degrees are in each angle of the triangle?

SOLUTION

It usually helps to draw a picture when solving geometry problems.

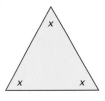

Equilateral Triangle

The sum of the angle measures of a triangle is 180°. Let x represent each angle measure. Since an equilateral triangle has three angles with equal measures, you can write the following equation:

$$3x = 180$$

If you divide each side by 3, the result is $x = 60$. Thus, each angle is 60°. Does this answer make sense?

LESSON ASSESSMENT

1 Explain how solving an equation is like using a pan balance.

2 How is the Division Property of Equality used to solve a multiplication equation?

3 Explain how to solve $ax = c$ for x.

4 How is the Property of Multiplying by 1 used in solving a multiplication equation?

5 Why is it necessary to check your answer to a problem after you solve the equation?

Practice and Problem Solving

Solve each equation for the given variable.

6. $8c = 64$

7. $8c = -64$

8. $-8c = 64$

9. $-8c = -64$

10. $5r = -27$

11. $12m = 80$

12. $15t = 140$

13. $-9p = 87$

14. $2.5y = 20$

15. $80 = -6z$

16. $-8 = -12b$

17. $7.5 = 2.5d$

18. $72 = -3.6d$

19. $-10q = -5$

20. $1.5e = 9$

21. $-0.3f = 9$

Solve each equation for the indicated variable. Give a reason for each step.

22. Solve for L in the equation $A = LW$.

23. Solve for i in the equation $V = ir$.

24. Solve for t in the equation $d = rt$.

25. Solve for p in the equation $i = prt$.

For each problem, write and solve an equation. Check your answer.

26. How many hours does it take a jet to fly 729 miles at a rate of 485 miles per hour?

27. A brick mason is building a patio in the shape of a parallelogram. The area of a parallelogram is the product of its base and height. Paul needs the patio to be 180 square feet. If the base of the patio is 12 feet, what is the height?

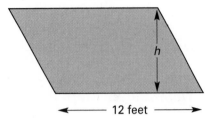

12 feet

28. The sum of the angles in a square is 360°. Each of the angles in a square has the same measure. Show that each angle of a square measures 90°.

Mixed Review

29. The distance from Tom's apartment to the library is 2.4 kilometers. How far is this distance in meters?

30. A carton of butter weighs 2.25 kilograms. What is this weight in grams?

31. A container holds 520 milliliters of water. How many liters does the glass hold?

32. Paul weighs 90,250 grams. What is Paul's weight in kilograms?

LESSON 4.2 THE MULTIPLICATION PROPERTY OF EQUALITY

Knowing how to work with equations will help you solve everyday problems as well as problems on the job. Sometimes you will use multiplication to solve these problems.

Solving Division Equations

ACTIVITY

The Multiplication Property of Equality

1 Start with the equation $\frac{x}{5} = 3$.

2 Multiply each side of the equation by 5.

3 What is the result?

4 Use a pan balance to explain what happened.

This Activity leads to another basic property of equality.

> **Multiplication Property of Equality**
> If each side of an equation is multiplied by the same number, the two sides of the equation stay equal or balanced.

An auto assembly plant has just started a night shift. The plant operates with teams of 6 people each. How many people are needed to work the night shift if there are 7 teams? The equation that models this problem is $\frac{t}{7} = 6$, where t is the total number of people on the night shift. To solve this problem, use the Multiplication Property of Equality to isolate the variable.

$$\frac{t}{7} = 6 \qquad \text{Given}$$

$$7 \cdot \frac{t}{7} = 7 \cdot 6 \qquad \text{Multiplication Property}$$

$$t = 42 \qquad \text{Simplify}$$

The auto assembly plant will need 42 people to work the night shift.

Use the Multiplication Property of Equality to solve the equation

$$\frac{m}{-8} = -9$$

EXAMPLE 1 Finding the Mass of an Object

You can find the density (D) of an object by dividing its mass (M) by its volume (V). The density of water is 1 gram per cubic centimeter. What is the mass of 25 cubic centimeters of water?

SOLUTION

You are given that $D = \frac{M}{V}$. If you multiply each side of the equation by V, the result is $DV = M$ or $M = DV$. Thus, $M = 25$ grams.

Reciprocals

If the product of two numbers is 1, the numbers are called **reciprocals** or **multiplicative inverses**. Since $6 \cdot \frac{1}{6} = 1$, 6 and $\frac{1}{6}$ are reciprocals. What is the reciprocal of $\frac{3}{2}$?

Reciprocal Property
The product of any number and its reciprocal is 1.

Critical Thinking Which numbers are their own reciprocals? Explain why.

In Lesson 4.1, you solved $5x = 15$ by using the Division Property. Now, using reciprocals, you can solve the same equation with the Multiplication Property.

$$5x = 15 \qquad \text{Given}$$

$$\frac{1}{5} \cdot 5x = \frac{1}{5} \cdot 15 \qquad \text{Multiplication Property}$$

$$x = 3 \qquad \text{Reciprocal Property; Simplify}$$
$$\left(\frac{1}{5} \cdot 5 = 1 \text{ and } \frac{1}{5} \cdot 15 = 3 \right)$$

The solution is 3. Check to make sure the solution is correct.

$$5(3) = 15$$
$$15 = 15$$

The purchase price (P) of a series EE Savings Bond is found by the formula $\frac{1}{2}F = P$, where F is the face value of the bond. Use the formula to find the face value of a savings bond purchased for $500.

The Multiplication Property and the Reciprocal Property are often used together to solve a multiplication equation.

EXAMPLE 2 Finding the Speed of a Car

A technician at an automobile test track is calibrating a new speedometer. She drives the car 45 miles in 45 minutes ($\frac{3}{4}$ hours). What is the speed of the car?

SOLUTION

To find the distance (d), multiply the rate (r) and the time (t). You can express this relationship as the equation

$$d = rt \quad \text{or} \quad rt = d$$

Now substitute the values for d and t into the equation and solve.

$rt = d$	Distance Formula
$r\left(\dfrac{3}{4}\right) = 45$	Given
$\dfrac{4}{3} \cdot \dfrac{3}{4}r = \dfrac{4}{3} \cdot 45$	Multiplication Property
$r = 60$	Reciprocal Property; Simplify

The speed of the car is 60 miles per hour. Does this answer make sense?

Rearranging Equations

When you solve an equation, it is often necessary to rearrange the numbers and variables. Two basic number properties allow for this rearrangement.

One basic property allows you to change the *order* of the numbers or variables in an equation.

Commutative Property of Multiplication

For all numbers a and b,

$ab = ba$

The Commutative Property states that the order in which you multiply two numbers has no affect on the product.

For example, $3 \cdot 4 = 4 \cdot 3$ and $2(5 \cdot 8) = (5 \cdot 8)2$.

The other basic property allows you to change the *grouping* of the numbers or variables in an equation.

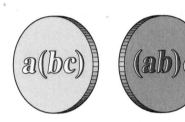

Associative Property of Multiplication

For all numbers a, b, and c,

$a(bc) = (ab)c$

The Associative Property states that the *grouping* of numbers you multiply has no affect on the product.

For example, $5(2 \cdot 8) = (5 \cdot 2)8$.

LESSON ASSESSMENT

Think and Discuss

1 How is the Multiplication Property used to solve a division equation? Give an example.

2 How do you find the reciprocal of a number?

3 How is the Multiplication Property used to solve a multiplication equation? Give an example.

4 How are the Commutative Property and the Associative Property used to rearrange the numbers and variables in an equation? Give an example of each.

Practice and Problem Solving

Find the reciprocal of each number.

5. 4   **6.** -6 **7.** $\dfrac{9}{10}$ **8.** $-2\dfrac{3}{4}$

Use the Multiplication Property to solve each equation. Check your answer.

9. $\dfrac{b}{12} = -7$

10. $8t = -128$

11. $125 = 5y$

12. $26 = \dfrac{m}{-2}$

13. $-9x = -66$

14. $-\dfrac{3}{5}a = 15$

15. $\dfrac{8}{5}k = -40$

16. $-36 = 1\dfrac{2}{3}d$

17. Solve the equation $A = LW$ for L. For W.

18. Write and solve an equation for this problem: Latisha has saved \$230 for a new compact stereo. So far, Latisha has $\dfrac{2}{3}$ of the amount she needs to buy the stereo. What is the cost of the stereo?

Mixed Review

Write each number in scientific notation.

19. 0.000056   **20.** 985,000   **21.** 143.95

From 3 AM until 8 AM, the temperature dropped two degrees an hour.

22. Write a multiplication expression to model the drop in temperature.

23. What is the change in temperature?

Twenty gallons of paint cost a painting contractor \$170.

24. Let a represent the cost of one gallon. Write a multiplication equation relating the cost of 20 gallons of paint to the total cost.

25. Use the equation you wrote to find the cost of one gallon of paint to the nearest cent.

LESSON 4.3 SOLVING PROPORTIONS AND PERCENT EQUATIONS

Ratio and Proportion

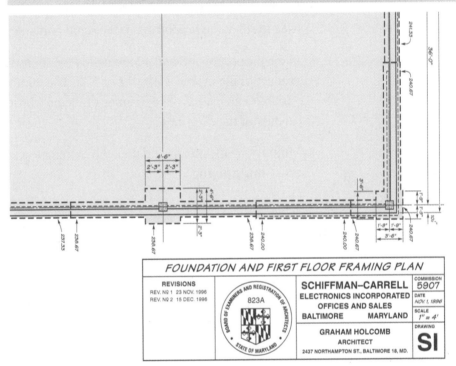

FOUNDATION AND FIRST FLOOR FRAMING PLAN

REVISIONS		COMMISSION 5907
REV. NO 1 23 NOV. 1996	SCHIFFMAN–CARRELL	DATE NOV. 1, 1996
REV. NO 2 15 DEC. 1996	ELECTRONICS INCORPORATED OFFICES AND SALES	SCALE 1" = 4'
823A	BALTIMORE MARYLAND	
	GRAHAM HOLCOMB ARCHITECT 2437 NORTHAMPTON ST., BALTIMORE 18, MD.	DRAWING SI

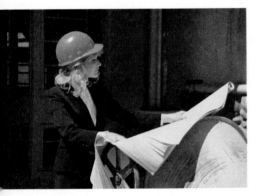

The scale drawing of an office building has a scale of 1 inch to 4 feet. This means that one inch on the drawing is the same as four feet in the actual office building. The scale is written as 1" : 4' or 1 : 4.

A **ratio** is a fraction that compares two quantities. Thus, you can also write the scale as $\frac{1}{4}$. However, it is better to keep the ratio in the same units. Since 4 feet equals 48 inches, this ratio can be written in one of three ways:

$$1" : 48" \qquad 1 : 48 \qquad \frac{1}{48}$$
$$\downarrow \qquad\qquad \downarrow \qquad\qquad \downarrow$$
$$\text{scale} \qquad\quad \text{ratio} \qquad\quad \text{fraction}$$

The length of the office building on the drawing is 12 inches. How can you find the full-scale length? Let s represent the full-scale length of the office building. First, form the ratio $\frac{12}{s}$.

The numerator represents the scale drawing length →12
The denominator represents the full-scale length ⟶ s

In this ratio, the units in the numerator and denominator are the same.

A **proportion** is an equation with a ratio on each side of the equal sign. Since the drawing is to scale, a proportion models the problem.

$$\frac{1}{48} = \frac{12}{s}$$

Critical Thinking   Show how the Multiplication Property can be used to solve the proportion in the scale drawing problem.

In the proportion, 48 and 12 are called the **means**. Thus, the product of the means is $48 \cdot 12$ or 576. The number 1 and the variable s are called the **extremes** of the proportion. What is the product of the extremes?

Multiplying the means and then the extremes is an example of **cross-multiplying**.

ACTIVITY	**Cross-Multiplication**

Find the product of the means. Then find the product of the extremes.

a. $\frac{3}{4} = \frac{6}{8}$ **b.** $\frac{4}{10} = \frac{6}{15}$ **c.** $\frac{-4}{8} = \frac{5}{-10}$ **d.** $\frac{10}{15} = \frac{4}{d}$

1 What do you notice about the results in **a, b,** and **c**?

2 Use your observation to solve the proportion in **d**.

Cross-multiplying is a method used to test whether two ratios are equal. For example, you can use cross-multiplication to show the two fractions, $\frac{3}{6}$ and $\frac{4}{8}$, are equal.

Find the product of the means: $3 \cdot 8 = 24$. Find the product of the extremes: $6 \cdot 4 = 24$. The cross-products are equal. Thus, the ratios are equal.

This leads to an important property for proportions.

Proportion Property
In a true proportion, the cross-products are equal.

The Proportion Property is a tool for finding the actual length of the office building.

$$\frac{1}{48} = \frac{12}{s}$$ Given

$$1 \cdot s = 12 \cdot 48$$ Proportion Property

$$s = 576$$ Property of Multiplying by 1

Thus, the actual length of the office building center is 576 inches or 48 feet.

Ongoing Assessment

Write and solve a proportion for this problem. A scale drawing for a rectangular tabletop measures 3 inches wide and 4 inches long. You want to make a table that is 5 feet wide. How long will it be? Show your work. Express your answer in feet and inches.

Percent

Percent means hundredths of a quantity. For example, 8% is just another way to write the ratio 8:100 or $\frac{8}{100}$. You can use proportions to change fractions or decimals to percents.

EXAMPLE 1 Changing Fractions to Percents

Change $\frac{5}{8}$ to a percent.

SOLUTION

Write a proportion $\dfrac{5}{8} = \dfrac{x}{100}$ $\left(\dfrac{x}{100} \text{ represents a } \% \right)$

and solve $8x = 500$ Proportion Property

$$x = 62.5$$

Thus, $\frac{5}{8}$ is equal to 62.5%.

Although proportions solve percent problems, it is usually easier to use a *percent equation*. One way to write a percent equation is

$$a\% \text{ of } b = c$$

In this form, a is the percent, b is the base, and c is the percentage.

Critical Thinking   Explain why the percent equation can also be written in the form:

$$\frac{a}{100} \cdot b = c$$

EXAMPLE 2 Using Percent to Find Sales

Tad earns a 6% commission on all of his sales. If Tad earns $150, what are his sales?

SOLUTION

Use a percent equation. The percent is 6, and the percentage is 150. Let b represent the base (Tad's sales). Write a percent equation.

$$\frac{6}{100} b = 150$$

Multiply each side by $\frac{100}{6}$.

$$\frac{100}{6} \cdot \frac{6}{100} b = \frac{100}{6} \cdot 150$$

$$b = 2500$$

Tad has sales of $2500.

WORKPLACE COMMUNICATION

How many gallons of liquid nitrogen will Cryogenics Supply Corporation deliver?

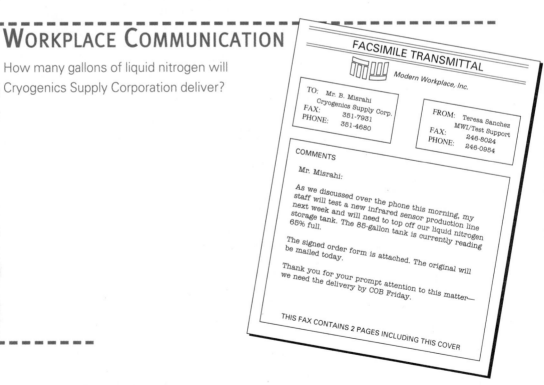

FACSIMILE TRANSMITTAL

Modern Workplace, Inc.

TO: Mr. B. Misrahi
       Cryogenics Supply Corp.
FAX: 351-7931
PHONE: 351-4680

FROM: Teresa Sanchez
         MWI/Test Support
FAX: 246-8024
PHONE: 246-0954

COMMENTS

Mr. Misrahi:

As we discussed over the phone this morning, my staff will test a new infrared sensor production line next week and will need to top off our liquid nitrogen storage tank. The 85-gallon tank is currently reading 65% full.

The signed order form is attached. The original will be mailed today.

Thank you for your prompt attention to this matter— we need the delivery by COB Friday.

THIS FAX CONTAINS 2 PAGES INCLUDING THIS COVER

LESSON ASSESSMENT

Think and Discuss

1 How are ratios and proportions related? Give an example.

2 Explain the meaning of the Proportion Property.

3 Explain how to use a proportion to change a fraction to a percent.

4 Explain how to use a percent equation to find what percent one number is of another.

Practice and Problem Solving

Solve each proportion.

5. $\dfrac{r}{5} = \dfrac{8}{10}$ **6.** $\dfrac{7}{3} = \dfrac{m}{9}$ **7.** $\dfrac{-6}{16} = \dfrac{15}{t}$ **8.** $\dfrac{-a}{12} = \dfrac{6}{9}$

Set up a proportion and solve. Round answers to the nearest tenth.

9. What is 25% of 120?

10. What percent of 10 is 6?

11. 64 is what percent of 256?

12. Find 12.5% of 72.

13. 6 is what percent of 9?

14. 9 is what percent of 6?

Set up a percent equation and solve. Round answers to the nearest tenth.

15. What percent of 300 is 13.5?

16. What is 300% of $20?

17. What is 0.5% of $120?

18. What is 15% of $225?

19. $2.50 is what percent of $10?

20. What percent of 12 is 2.4?

For each problem, set up a proportion or percent equation and solve.

21. Keesha bought a sweater for $50. If the sales tax rate is 8.5%, find the total amount Keesha paid the cashier.

22. Sami borrowed $1200 for one year. If Sami paid $90 simple interest, what was the interest rate?

23. Tom pays 6.2% of his wages for social security. If Tom paid $186 in social security taxes, what was his income?

24. Rhonda buys her plumbing supplies at wholesale. She buys pipe for $4/ft and sells it for $5/ft. What is the percent mark-up?

Simplify.
25. $(4.2 \bullet 10^{-3})(5.9 \bullet 10^{-2})$

26. Water is stored in a rectangular container that is 8 inches by 12 inches by 4 inches. The water is transferred to a spherical container with a radius of 10 inches. Is the spherical container large enough to hold the water? By how much is it too small or too large?

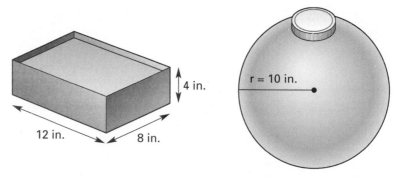

27. Evaluate $-8c + 2d \div -4$ when c is 5 and d is -6.

28. Solve the equation $mx = b$ for x. Show your steps. What properties did you use?

LESSON 4.4 THE ADDITION PROPERTY OF EQUALITY

Subtraction Equations

The Blueberry Baking Company has announced a 48¢ reduction in price for its cookie products. Cookies that sold for $1.59 last week will now cost $1.59 − $0.48, or $1.11. Use your calculator to check whether this computation is correct.

You can model the drop in cookie price by an equation. Represent the original cookie price by p. Represent the new cookie price by n. The equation that models the relationship between the old and new cookie prices is the subtraction equation

$$n = p - 0.48$$

After the sale, the cookies return to their original price. If you add 0.48 to the right side of the equation, it will reverse the effect of subtracting 0.48. Thus, adding 0.48 will isolate p. However, to maintain the balance of the equation, you must also add 0.48 to the left side.

Addition Property of Equality

If the same number is added to each side of an equation, both sides of the equation stay equally balanced.

The Addition Property says that if you add 0.48 to the right side of the equation, you must also add 0.48 to the left side. Here are the steps that isolate p.

$n = p - 0.48$	Given
$n + 0.48 = p - 0.48 + 0.48$	The Addition Property
$n + 0.48 = p + 0$	Simplify
$n + 0.48 = p$	

Using the Addition Property, you have isolated p on one side of the equation while maintaining the "balance" of the equation.

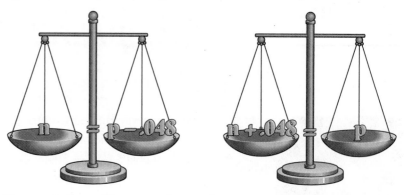

The pan balance illustrates the results.

Addition Equations

Now suppose the Blueberry Baking Company increases prices by $0.48 over the original price. The equation

$$n = p + 0.48$$

models this increase. In this equation, the variables p and n have the same meaning as before. Except this is now an addition equation. If you want to isolate p in this equation, you can subtract 0.48 from each side of the equation. When the same value is subtracted from each side of the equation, you are using the Subtraction Property of Equality.

Subtraction Property of Equality
If the same number is subtracted from each side of an equation, both sides of the equation stay equally balanced.

Before you use the Subtraction Property to solve the equation

$$n = p + 0.48$$

for p, another property is needed. Do you remember using this property earlier in this lesson?

Property of Adding Zero
For any number a,

$a + 0 = a$ and $0 + a = a$

Now you can use the Subtraction Property to solve for p.

$n = p + 0.48$	Given
$n - 0.48 = p + 0.48 - 0.48$	Subtraction Property
$n - 0.48 = p$	Explain why

EXAMPLE 1 Using Sales Tax

The total price (p) of a car is equal to the sum of the sale price (s) and the sales tax (t). Write an addition equation and solve for the sale price.

SOLUTION

One way to write the addition equation is $s + t = p$. If you subtract t from each side of the equation, the result is

$$s = p - t$$

Ongoing Assessment

Solve the following equations for the indicated variable.

1. $n = p - 0.64$ for p

2. $y = x - 12$ for x

3. $k = m + 25$ for m

Sometimes you must rearrange the numbers and variables in an addition or subtraction equation. Remember the two properties you used in Lesson 4.2 to rearrange numbers and variables in a multiplication equation. The Commutative and Associative Properties are also true for addition.

Commutative Property of Addition
For all numbers a and b,

$$a + b = b + a$$

For example, $3 + 7 = 7 + 3$.

Associative Property of Addition
For all numbers a, b, and c,

$$a + (b + c) = (a + b) + c$$

For example, $6 + (4 + 8) = (6 + 4) + 8$.

Critical Thinking   Which basic property is illustrated?

a. $5(3 + 4) = (3 + 4)5$ b. $(1 + 2) + 3 = 3 + (1 + 2)$

EXAMPLE 2 Rearranging Equations

Solve the equation $15 + x = -32$.

SOLUTION

The Commutative Property states that you can add in any order. To rearrange the left side of the equation, replace $15 + x$ with $x + 15$.

$$x + 15 = -32$$

Now subtract 15 from each side,

$$x + 15 - 15 = -32 - 15$$
$$x = -47$$

The solution is -47.

The Associative Property states that you can regroup the numbers or variables in an equation. For example, to solve $(r + 5) + 10 = 25$, the Associative Property allows you to regroup to obtain $r + (5 + 10) = 25$. These properties are used together to rearrange the terms in an equation. This will simplify your work when solving problems in more complex situations.

EXAMPLE 3 Rearranging to Solve an Equation

Solve the equation $(-6 + n) + (3 - 2n) = 0$.

You can use the definition of subtraction and at the same time rearrange the left side to rewrite the equation.

$$3 + (-6 - n) = 0$$ Given

$$-n + (3 - 6) = 0$$ Rearrange the terms using the Commutative Property and the Associative Property

$$-n - 3 = 0$$ Simplify

$$-n = 3$$ Addition Property

There are two ways to solve the equation $-n = 3$.

1. Think of the negative sign as an opposite sign. If the opposite of n is 3, n must be -3.

2. Think of $-n$ as a short way of writing $-1 \cdot n$. Multiply each side by -1. The result is $n = -3$.

By either method, the solution is -3. Check: $(-6 - 3) + (3 + 6) = 0$
$$-9 + 9 = 0$$

The solution to Example 3 leads to an important definition. You can find the opposite of a number by multiplying the number by -1.

Multiplying by -1
For any number a,

$$-a = -1a$$

EXAMPLE 4

Solve the equation $17 = 9 - n$.

SOLUTION

Subtract 9 from each side of the equation. The result is

$$-n = 8$$

The solution is -8 because $(-1)(-8) = 8$ and $-1(-n) = n$.

LESSON ASSESSMENT

Think and Discuss

1 Compare the Addition Property of Equality and the Subtraction Property of Equality. How are they alike and different?

2 How is the Property of Adding Zero used to solve an addition or subtraction equation?

3 Explain how to solve the equation $a + b = c$ for b.

Practice and Problem Solving

Solve each equation. Show each step.

4. $y + 5 = 9$ **5.** $22 = t + 18$ **6.** $q - 9 = -12$

7. $36 = c - 27$ **8.** $-14 = k + 36$ **9.** $18 + a = 17$

10. $-k = -8$ **11.** $18 - w = 20$ **12.** $36 = -15 - e$

Solve each equation. Check your answer.

13. $4.6 = z + 5.4$ **14.** $p - (-12) = 20$ **15.** $-1.9 = s - 5.1$

16. $f - 3.5 = -2.8$ **17.** $10 = 5.6 + n$ **18.** $-8.5 - d = -3$

19. $5.3 + x = -7.4$ **20.** $6 - x = -15.3$ **21.** $x - 2.78 = -5.9$

22. The formula for profit (p) is cost (c) minus overhead (o). Write an equation for profit and solve it for cost.

23. The equation $556 + n = 580$ represents the change in the number of computer workstations in a corporation over one year. Solve the equation for n to find the change in the number of workstations.

Ross wrote a check for $124 and received a notice from the bank stating that he was $51 overdrawn. His account now shows a negative balance.

24. Let *a* represent the original amount in Ross's account. Write an equation to model the bank's transaction.

25. Find the original amount in Ross's account.

Mixed Review

On three successive carries of the football, Todd lost 5 yards, gained 8 yards, and lost 3 yards. He started at the 25-yard line.

26. Use integers to show each carry.

27. On what yard line will the team start the next play?

On a vacation trip, Maria drove a total of 978 miles in 6 days.

28. Write an equation to model the number of miles Maria drove each day.

29. Solve the equation.

LESSON 4.5 SOLVING MULTISTEP EQUATIONS

Equations with One Variable

Many equations contain several terms. Often you must use more than one of the properties of equality to solve these equations. In the process of solving an equation with one variable, isolate the variable and simplify the constants on the other side of the equation.

EXAMPLE Solving a Linear Equation

Solve $3g - 7 = 11$.

SOLUTION

Start by identifying the variable. In this equation, it is the letter g. Then begin the process of isolating g.

Look at the equation $3g - 7 = 11$ and ask yourself these questions.

- How can I isolate the expression containing g from the other expressions on the same side of the equation?

- Once g and its coefficient are isolated, how can I isolate g on one side of the equation— that is, what is the next thing I must do to free g from $3g$?

As you carry out this process, always remember that whatever you do to one side of an equation—you must do the same to the other side. Remember to keep the equation balanced.

$$3g - 7 = 11 \qquad \text{Given}$$
$$3g - 7 + 7 = 11 + 7 \qquad \text{Addition Property}$$
$$3g = 18 \qquad \text{Simplify}$$
$$\frac{3g}{3} = \frac{18}{3} \qquad \text{Division Property}$$
$$g = 6 \qquad \text{Simplify}$$

Finding a solution is only half of the procedure when you solve an equation. You must also check to make sure the value you found is correct. To check your work, substitute your solution in the original

equation. Since you are not sure that the two sides are really equal (remember you are checking your answer), write a question mark over the equal sign.

$$3g - 7 = 11 \qquad \text{Original equation}$$

$$3(6) - 7 \overset{?}{=} 11 \qquad \text{Substitute the value 6 for the variable } g$$

$$18 - 7 \overset{?}{=} 11 \qquad \text{Simplify by doing the arithmetic}$$

$$11 = 11 \qquad \text{Verify the equality}$$

Ongoing Assessment

Solve each equation. Show each step. Check your answers.

a. $4a + 5 = -15$ **b.** $8 - 3a = 9$ **c.** $\dfrac{2}{3}y - 1 = -13$

Using the Distributive Property

Recall how you used an equation to solve John's problem in Lesson 4.1. John had just started a job that paid an hourly rate of $5.35 per hour. His work schedule was 15 hours during the week and 8 hours on Saturday. This equation models John's weekly pay:

$$p = 5.35(15 + 8) = 123.05$$

John decides he needs to earn $150 per week while still working the same number of hours. Thus, he needs a new hourly rate r. He needs to replace p by 150 and 5.35 by r.

$$150 = r(15 + 8)$$

Solving this equation shows that John needs to make over $6 per hour. Since John probably will not make this rate in the near future, he must find a way to increase his weekday hours. To find the number of weekday hours, start with the original equation for weekly pay.

$$p = 5.35(15 + 8)$$

Again replace p with 150. Since 15 (the number of weekday hours) is going to change, replace 15 by the variable e. The 8 hours of weekend time remains the same. The new equation is

$$150 = 5.35(e + 8)$$

There are two ways to solve for e. One way is to solve for $(e + 8)$ by dividing both sides by 5.35 (Division Property of Equality). Then subtract 8 from both sides (Subtraction Property).

$150 = 5.35(e + 8)$	Given
$\dfrac{150}{5.35} = \dfrac{5.35(e + 8)}{5.35}$	Division Property
$28.04 = e + 8$	Simplify
$28.04 - 8 = e + 8 - 8$	Subtraction Property
$20.04 = e$	Simplify

Check to see if 20.04 is the correct solution. John must work just over 20 hours during the weekdays to earn at least $150 per week.

The other method for solving John's equation for e uses the Distributive Property. The following Activity will give you an opportunity to see how this property is used to isolate the variable e.

ACTIVITY 1   Using the Distributive Property First

Copy these steps on your paper. Give reasons for each step.

	Steps	Reasons
1	$150 = 5.35 (e + 8)$	Given
2	$150 = 5.35e + 5.35 \cdot 8$	?
3	$150 = 5.35e + 42.8$	?
4	$150 - 42.8 = 5.35e + 42.8 - 42.8$	?
5	$107.2 = 5.35e$	?
6	$\dfrac{107.2}{5.35} = \dfrac{5.35e}{5.35}$	?
7	$20.04 = e$	?

Critical Thinking   How do the two methods for solving the equation $150 = 5.35 (e + 8)$ compare? Which one do you prefer? Why?

Ongoing Assessment

Use two methods to solve the equation $5(t - a) = m$ for t. Show your steps and give a reason for each one.

You can write a general equation that describes John's salary situation. Let p be the weekly pay. Let r be the hourly rate. Let e be the hours worked weekday evenings. Let s be the hours worked on Saturdays. Then, the general equation is

$$p = r(e + s)$$

You can solve this equation for any of the variables. Complete the Activity to solve for r in terms of p, e, and s.

ACTIVITY 2   Using the Division Property First

Give reasons for each step.

	Steps	Reasons
1	$p = r(e + s)$	?
2	$\dfrac{p}{(e + s)} = \dfrac{r(e+s)}{(e+s)}$	?
3	$\dfrac{p}{(e + s)} = r$	?

The result of the activity is that r is now isolated on one side of the equation. Since you used only basic properties, the equation should be "balanced." To check your work, find numbers that make the original equation true. Then substitute these same numbers into the new equation and see if both sides are still equal.

WORKPLACE COMMUNICATION

Explain how you can use the ideas in Chapter 4 to reach this conclusion.

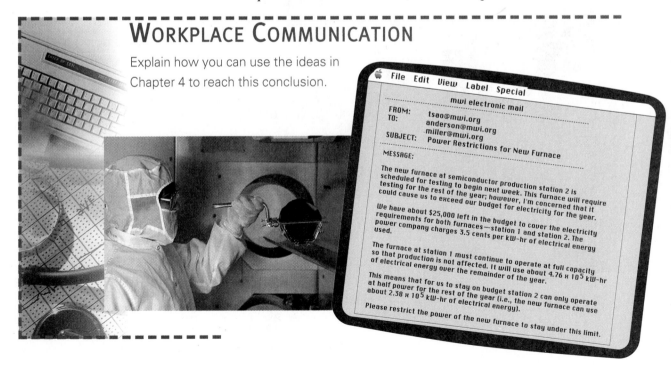

File   Edit   View   Label   Special

mwi electronic mail

FROM: tsao@mwi.org
TO: anderson@mwi.org
         miller@mwi.org
SUBJECT: Power Restrictions for New Furnace

MESSAGE:

The new furnace at semiconductor production station 2 is scheduled for testing to begin next week. This furnace will require testing for the rest of the year; however, I'm concerned that it could cause us to exceed our budget for electricity for the year.

We have about $25,000 left in the budget to cover the electricity requirements for both furnaces—station 1 and station 2. The power company charges 3.5 cents per kW-hr of electrical energy used.

The furnace at station 1 must continue to operate at full capacity so that production is not affected. It will use about 4.76×10^5 kW-hr of electrical energy over the remainder of the year.

This means that for us to stay on budget station 2 can only operate at half power for the rest of the year (i.e., the new furnace can use about 2.38×10^5 kW-hr of electrical energy).

Please restrict the power of the new furnace to stay under this limit.

LESSON ASSESSMENT

Think and Discuss

1 Explain how to solve $2(a + b) = c$ for a in two different ways.

2 Why is it necessary to check the solution to an equation?

Practice and Problem Solving

Solve each equation. Show each step. Check your answer.

3. $2x + 15 = 29$ **4.** $16 = 4t - 20$ **5.** $7 + 3r = -11$

6. $-5x + 9 = 4$ **7.** $-6s - 12 = 24$ **8.** $3(x + 2) = -5$

Solve each equation. Check your answer.

9. $2.5(4 + b) = -20$ **10.** $3.7 = 1.8f - 3.9$

11. $-28 = 3.6(m - 6)$ **12.** $\frac{2}{3}c + 5 = 13$

13. $\frac{3}{4}(8 + 2y) = 3$ **14.** $\frac{5b + 9}{4} = -4$

15. Solve the equation $ax + b = c$ for x.

16. Solve the equation $a(x + b) = c$ for x.

The formula for the perimeter of a rectangle is $P = 2L + 2W$. P represents the perimeter, L the length, and W the width. Substitute the dimensions into the formula and solve the equation for the missing variable. (All measurements are in feet.)

17. length is 15; width is 12

18. perimeter is 60; length is 12

19. perimeter is 52; length is 15

20. perimeter is 36; width is 3

21. Solve the perimeter formula for L.

22. Rewrite the perimeter formula using the Distributive Property.

23. The sum of Lin's three test scores for the grading period is 264. There is one test remaining, and Lin would like to average 90 on his test scores. Lin uses the equation $90 = \frac{264 + s}{4}$ to find the fourth test score he must make. What is the test score?

24. A 60-liter solution is 30% acid. The formula $180 = 2(n + 60)$ can be used to find the number of liters of water (n) that must be added to make a 20% solution. How many liters must be added?

25. Natasha is running a "20% off" sale at her clothing shop. On one jacket, she is offering a savings of $25. What does Natasha usually charge for the jacket?

Mixed Review

Simplify
26. $-35 - (6 + 4 \bullet -5)$ **27.** $(2.6 \bullet 10^{-5})(5.12 \bullet 10^7)$

The formula to find the volume (V) of a cylinder with radius (r) and height (h) is $V = \pi r^2 h$.

28. The Cup Company produces coffee mugs in the shape of cylinders. The mugs have a radius of 4 centimeters and a height of 9 centimeters. What volume of coffee will one of the mugs hold?

29. A restaurant wants to order mugs that hold 706.5 cubic centimeters of soup each. For stacking purposes, the cups must have a 10 centimeter diameter. What height should the cups be?

30. Solve the equation $\frac{c}{d} = \frac{m}{n}$ for d.

LESSON 4.6 EQUATIONS WITH VARIABLES ON BOTH SIDES

An electronics manufacturing plant produces telephones at the rate of 600 per day. A customer survey indicates that the demand for phones with built-in answering machines is twice as great as for those without them. If you are determining the production quota for a day, how many phones with answering machines would you schedule for production? Model the problem with an equation.

Let x represent the number of phones to be manufactured without the answering machines. Then $2x$ represents the number of phones with answering machines. The total number of phones must be 600.

$$\begin{array}{ccccc} \text{Phones without} & + & \text{phones with} & = & 600 \\ \text{machines} & & \text{machines} & & \text{phones} \\[4pt] x & + & 2x & = & 600 \end{array}$$

To find the number of each type of phone, you need to solve the equation

$$x + 2x = 600$$

To see how to simplify the expression $x + 2x$, you can use a rectangular shaped block called an Algeblock.

This Algeblock has a height of one unit and a length of x units. Remember that the coefficient of x is understood to be one.

That is,

$$1x = x$$

$$1x \quad + \quad 2x \quad = \quad 3x$$

Now use the Division Property to solve for x. If $3x = 600$, then $x = 200$ and $2x = 400$.

Thus, you will schedule the production of 200 phones without answering machines and 400 with answering machines. Does this answer make sense?

If all the terms in an expression have the same variable, the terms are called **like terms.** You solved the telephone problem by adding like terms using Algeblocks as a model. You can also add like terms by using the Distributive Property.

$$1x + 2x = (1 + 2)x$$
$$= 3x$$

EXAMPLE 1 Simplifying Expressions

Simplify the expression $3a + 2b + 4a + 4b$.

SOLUTION

$$3a + 2b + 4a + 5b = 3a + 4a + 2b + 4b \qquad \text{Rearrange terms}$$
$$= 7a + 6b \qquad \text{Add like terms}$$

You can also subtract like terms.

ACTIVITY   **Subtracting Like Terms**

1 Use a drawing of Algeblocks to simplify $5x - 3x$.

2 Use the definition of subtraction and the Distributive Property to simplify $5x - 3x$.

3 Explain why $6x - x$ is $5x$.

EXAMPLE 2 Equal Compensation

A salesperson in a stereo store is given a choice of two different compensation plans. One plan offers a weekly salary of $250 plus a commission of $25 for each stereo sold. The other plan offers no salary but pays $50 commission on each stereo sold. How many stereos must the salesperson sell to make the same amount of money under both plans?

Let x represent the number of stereos sold in one week.
Plan 1 offers a \$25 commission ($25x$) plus \$250 salary.
Plan 2 offers a \$50 commission ($50x$).

Find when plan 1 ($25x + 250$) = plan 2 ($50x$).

$25x + 250 = 50x$	Given
$250 = 50x - 25x$	Subtraction Property
$250 = 25x$	Subtract like terms
$10 = x$	Division Property

Since $25(10) + 250 = 500$ and $50(10) = 500$, selling 10 stereos a week results in the same compensation.

Critical Thinking   Solve the equation $5 - 2x + 3 = 4x + 8 - 6x$. What does the solution tell you about the equation?

CULTURAL CONNECTION

In 1858, Henry Rhind bought an ancient Egyptian papyrus that was written by the Egyptian scribe Ahmes. The Ahmes, or Rhind Papyrus as it is known, contains 85 problems copied by Ahmes from earlier writings. One of the problems illustrates how the Egyptians used a method now called "guess and check" to solve an equation such as $5x + 3 = 28$. Here is how it works.

Guess a solution and substitute to see if you are correct. Try 10.

Since $5(10) + 3$ is 53, the guess is too high. Try 2.

$5(2) + 3$ is 13. This guess is too low. The next guess might be halfway between the high and the low. Try it and see if it works. Do you see how the Egyptians eventually found the right number? Use the guess and check method to solve these two equations.

1. $6b - 9 = 30$ **2.** $9w + 18 = 5w - 4$

3. What are the advantages and disadvantages of the guess and check method?

LESSON ASSESSMENT

1 Explain how to use Algeblocks to add $7x + 3x$.

2 Explain how to use the Distributive Property to solve $7x + 3x = 20$.

3 Why can you add like terms?

4 Explain how to solve $3m - 2 = 5 - 4m$.

Practice and Problem Solving

Solve each equation. Show each step.

5. $3a + 9 + 4a = 15$

6. $4m - 8m - 10 = -24$

7. $5(n + 3) + 2n = -10$

8. $2(d - 3) - 5d = 18$

9. $w + 6 = w + 5$

10. $3r + 12 = 5r + 15$

Solve and check your solution for each equation.

11. $16 - p = 3p + 8 - 5p$

12. $5(y - 1) = -y$

13. $b - 8 = 4(2b + 3)$

14. $-4(3 + c) = 6(3 - c)$

15. $\frac{3}{4}(12x + 8) = 50x$

16. $\frac{t + 1}{9} = \frac{t}{-3}$

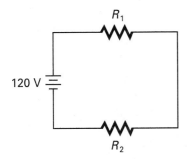

An electronics technician must connect two resistors in series with a 120-volt power supply. The total voltage across two resistors connected in series is the sum of the voltage across each resistor. The voltage across the second resistor is 60 volts greater than the voltage across the first resistor. Let x represent the voltage across the first resistor.

17. Write an expression for the voltage across the second resistor in terms of x.

18. Write an equation for the sum of the voltages across the resistors.

19. Find the voltage across each resistor.

A contractor is building a 4-foot circular sidewalk around a flower bed. The outside circumference of the sidewalk is 1.5 times greater than the circumference of the flower bed. Let x represent the radius of the flower bed.

20. Draw a picture to model the problem.

21. Write an equation that equates the outside circumference of the sidewalk to 1.5 times the circumference of the flower bed.

22. Solve the equation for the radius of the flower bed.

23. A manufacturer can produce 350 riding mowers per day. To meet the needs of distributors, the company must produce four times as many 38-inch mowers as 42-inch mowers. How many of each type of mower should be manufactured each day?

A person is using CAD software to draw rectangles. The software is programmed to produce rectangles as shown for each value of x.

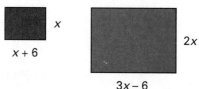

24. What value for x will produce rectangles with equal perimeters?

25. What are the dimensions of each of the rectangles in Exercise 24?

Mixed Review

The Marshall City Safety Department recorded the number of cars passing the intersection of Main and Jackson between 8 AM and 10 AM for five consecutive weekdays.

124   361   243   210   282

26. Find the average number of cars passing the intersection.

27. Use 250 as a benchmark. Subtract the benchmark from each recorded number. Write this deviation between the benchmark and each recorded number as an integer.

28. Find the average of the deviations. Add this number to the benchmark. Compare this sum to the average.

LESSON 4.7 EQUATIONS ON THE JOB

Solving an equation is a step-by-step process. However, finding a correct equation to model a problem is sometimes a difficult process. When you read a problem, try to understand the problem so that it makes sense to you. Follow these steps to help you understand the problem.

1. Look at all the words in the problem and make sure you understand what each word means. If you are not sure, ask someone or look up the word in a reference book, such as a dictionary. You must know what the words mean before you can begin to solve the problem.

2. After you understand the words, try to understand the situation. Picture in your mind the situation that the words describe. Draw a rough sketch of the situation.

3. Form a sentence (either in your mind or on your paper) that begins with what you are trying to find and continues with the verb *is*. For example, if you need to find the distance around a park, you might use this sentence: "The distance around the park *is* the sum of the lengths of all of the sides."

4. Rewrite your sentence in the form of a mathematical sentence or equation. Look for words in the problem that have specific mathematical meanings.

Explain how these steps are used in the following Activities.

ACTIVITY 1   A Soft Drink Problem

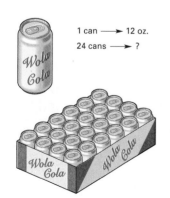

A catering company uses a 24-can case of 12 fluid-ounce drinks to make ice cream sodas for a party. The caterer will adjust the other ingredients of the ice cream sodas based on the number of ounces of soft drink that are used. How many ounces are in a case of soft drinks?

1 Are all the words in the problem understandable? Does the drawing help you understand the problem better?

2 You know the volume in one can of soft drink. You want to find the volume in a case of 24 such cans. Complete this sentence: "The weight of the soft drink in a case of 24 cans is . . . ."

3 Write an equation that models the problem.

ACTIVITY 2   Cutting a Metal Pipe

A plumber needs a 12.5 inch length of copper pipe. She has one piece that is 20.8 inches long. How much pipe must be cut off to leave a piece that is 12.5 inches long?

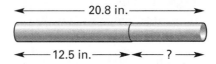

To "cut off" means to subtract. For this problem, use the sketch of the pipe to show how much to cut and how much to leave. You might begin your sentence in this way: "The piece cut off is . . . ." Finish the sentence and write an equation.

ACTIVITY 3   A Family Budget

The family budget has 10% of the take-home pay assigned to transportation. If the monthly take-home pay is $1200, how much is in the budget for transportation?

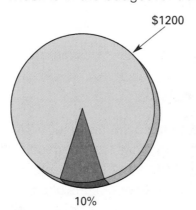

Do you know the meaning of the key words (budget, take-home pay, and transportation) in the problem? You might begin your sentence in this way: "The amount budgeted for transportation is . . . ." Finish the sentence and write an equation.

Compare your equations to the following ones. Often there is more than one correct way to write the equation for a problem. Each of the following equations shows just one of the possibilities.

1. $w = 24$ cans • 12 ounces per can, or $w = 24 • 12 = 288$ oz

2. $l = 20.8$ inches $- 12.5$ inches, or $l = 20.8 - 12.5 = 8.3$ in.

3. $t = 10\%$ of $1200, or $t = 0.10 • \$1200 = \$120$

When you use equations to solve problems on the job, the equations are not always written out for you. You may first have to restate the problem in equation form and then solve the equation.

Suppose the personnel manager of a large department store has just hired you. He explains that your weekly salary is $150. In addition, you will receive 5% of the total value of the sales you make during the week. How can you predict your gross pay for any week?

Remember that it is helpful to picture the problem as you begin to work toward a solution. In this case, you can picture your pay for a week as two stacks of money. The first stack is 150 dollar bills (or $150). The second stack represents 5% of your total sales of the week. The sum of the two stacks (in dollars) is your weekly pay.

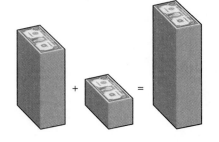

Now you can put your picture into words. Your weekly salary is $150 plus 5% of your total sales for the week. In equation form, this becomes

$$W = 150 + 0.05S$$

Here, W represents your weekly salary in dollars and S your total weekly sales in dollars. Notice that 5% is written as "0.05," the decimal form that your calculator understands. You can also rewrite your equation as

$$W = 0.05S + 150$$

5% =

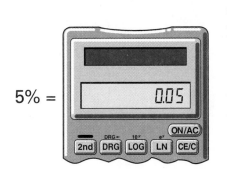

Suppose your total sales during the first week came to $1000. Substitute $1000 for S, and solve for W.

$$W = 0.05 (1000) + 150$$

$$W = 50 + 150$$

$$W = 200$$

The solution is $200. First check to make sure the arithmetic is correct. Then analyze the solution in terms of the situation. In this problem, 5% means $5 for each $100 you sell. It makes sense that for $1000, you will get ten times as much, or $50. Thus, your total salary for this week is $50 + $150, or $200.

Use the equation $W = 0.05S + 150$ and complete the table.

W	S
200	1000
?	500
?	200
?	100

Write a sentence that describes the relationship between W and S.

Ongoing Assessment

What if your salary for the week is $500? What is the amount of sales you made for the week? The same formula models this situation.

$$W = 0.05S + 150$$

Substitute $500 for W and solve for S. Analyze your solution in terms of the situation.

Now you are ready to write and solve the equations needed in the Math Lab Activities and Math Applications.

MATH LAB

Equipment Calculator
Soup can
1 pound coffee can
Wooden dowel about 1 inch in diameter
Vernier caliper
Cloth tape measure

Problem Statement

You will work with the formula for the circumference of a circle. Let C represent the circumference of a circle. Let d represent the diameter. Use 3.1416 as the approximate value of π.

$C = \pi d$ (this is Equation 1)

You will form new equations by solving the equation for d and for π. Then, you will check your work by using measured amounts to see if your equations balance.

Procedure

a Measure the circumference of the coffee can, soup can, and wooden dowel. Record your results as the first three entries under the "Measured Left Side" column.

Equation	Left Side	Right Side	Object	Measured Left Side	Calculated Right Side
Equation 1	C	πd	Coffee can		
			Soup can		
			Dowel		
Equation 2	d		Coffee can		
			Soup can		
			Dowel		
Equation 3	π		Coffee can	3.1416	
			Soup can	3.1416	
			Dowel	3.1416	

b Measure the diameter of each object with the vernier calipers. Complete the "Measured Left Side" column by entering these measurements.

c Solve Equation 1 for *d*. Call this Equation 2. Record the right side of this equation in the "Right Side" column.

d Solve Equation 1 for π. Call this Equation 3. Record the right side of this equation in the "Right Side" column.

e Calculate and record the values for the right side of Equations 1, 2, and 3 using measured values for circumferences and diameters.

f Are both sides of all your equations equal? If the answer is yes, you have solved all the equations correctly. If not, check to be sure you correctly used the basic properties for solving equations correctly.

g Explain why the values might not be exactly the same even though you solved each equation correctly.

Activity 2: Creating Equations

Equipment   Ball of string (over 400 feet long)
Masking tape
Tape measure
Calculator

Problem Statement

You will create an equation that determines the perimeter of an irregularly shaped area. You will then form several new equations by solving the perimeter equation for each of its variables. Finally, you will check the balance of these new equations.

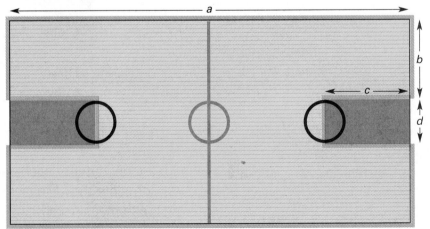

Procedure

a Measure the lengths of the irregular sections of the basketball court represented by *a*, *b*, *c*, and *d.* Record each value in Table 1.

Side	Length
a	
b	
c	
d	

Table 1.   Perimeter =

b Devise a way to use the string, tape, and tape measure to measure the highlighted perimeter in the illustration.

c Measure and record this perimeter in Table 1.

d Using the variables *a*, *b*, *c*, and *d*, create an equation that models the highlighted perimeter of the court. Record the right side of this equation in the first row of Table 2 under the "Right Side" column.

Left Side of the equation	Right Side of the equation	Measured Left Side	Calculated Right Side
p			
a			
b			
c			
d			

Table 2.

e Enter the measured value of the perimeter from Table 1 in the column labeled "Measured Left Side." Also calculate the value of the right side of the perimeter equation using the measures in Table 1. Record the result of your calculation in the first row of the "Calculated Right Side" column of Table 2.

f Solve the perimeter equation for the variable *a*. Record the right side of this new equation in the second row of the "Right Side" column.

g Complete the second row of Table 2 by using the measured values from Table 1.

h Complete Table 2 by repeating Steps **f** and **g** for the variables *b, c,* and *d.*

i What conclusions can you make about your calculations? What does the calculation tell you about the equations you created?

Activity 3: Solving Equations

Equipment   Algeblocks
                Algeblocks mat

If Algeblocks are not available, you can make paper or cardboard tiles in the dimensions shown here. Color half of each type of tile yellow and the other half green. Also make a mat like the one shown.

unit tile   *x*-tile

mat

Problem Statement

You will use Algeblocks or tiles to solve a simple equation.

Procedure

a Model the equation $3x - 2 = 7$ by placing the Algeblocks on your mat in this position:

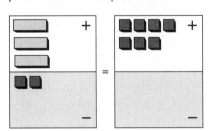

b Add two units to each side of the equal sign.

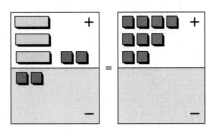

c Remove the two units on the positive section and the two units on the negative section of the left mat. What mathematical property is applied in this step?

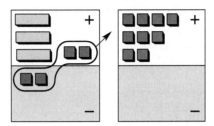

d Divide both sides into three groups. How many units are in each group?

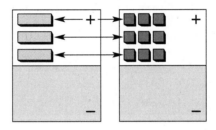

How does this show that 3 is the solution to $3x - 2 = 7$? Check your solution. Does it make the original equation true?

e Use Algeblocks to solve $5x + 3 = 2x - 3$. Record each step by drawing a picture. Write a summary of your findings.

MATH APPLICATIONS

The applications that follow are like the ones you will encounter in many workplaces. Use the mathematics you have learned in this chapter to solve the problems.

Wherever possible, use your calculator to solve the problems that require numerical answers.

Social security tax is deducted from your paycheck using the formula

$$T = P \times G$$

T is the tax amount to be deducted.
P is the decimal percentage rate for the tax.
G is your gross pay.

FOLD AND REMOVE					FOLD AND REMOVE

	RATE	HRS	AMT	YTD
MODERN WORKPLACE, INC. 652 INDUSTRY WAY GREENVILLE, CT 43121	7.54	40	$301.60	$301.60

ADJUSTMENTS	AMOUNT	YTD
$^\circ_c$ INDICATES OTHER COMP		

	AMT	YTD
JANE SMITH		
468 MAPLE STREET		
GREENVILLE, CT 43121	**TOTAL WAGES** $301.60	$301.60
	SOC SEC 22.65	22.65
	MEDICARE 4.37	4.37
TAX FILING STATUS PAY PERIOD	FEDERAL 49.40	49.40
FEDERAL S 01 01-07-97	CT 20.67	20.67
CT S 01		
CHECK DATE		
01-13-97		
CHECK NBR		
10359	**NET PAY** $204.51	$204.51

1 Write a sentence that describes how to compute the tax deducted from your paycheck.

2 Rewrite the formula isolating the variable *P*. In other words, write a formula for determining the value of *P*.

3 Suppose your paycheck stub shows gross pay of $301.60 and a tax deduction of $22.65. Use your formula to determine the percentage rate used for computing your deduction. Round the answer to two decimal places.

When a lever is used to lift or balance weights, as occurs in common patient scales seen in doctor offices, the sizes and positions of the weights are governed by the equation

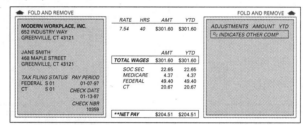

$$F_1 D_1 = F_2 D_2$$

F₁ and *F₂* are the weights (forces), and
D_1 and D_2 are the distances from the fulcrum.

4 You have a scale with a movable weight (F_2) that is 500 grams. The patient stands on a platform that is 0.2 cm (D_1) from the fulcrum. Write a formula you could use to compute the patient's weight (F_1) if the 500-gram weight is balanced at D_2 centimeters from the fulcrum.

5 What is the weight of the person on the scale if the scale balances when you place the 500-gram weight 18 centimeters from the fulcrum?

BUSINESS & MARKETING

It costs a recording company $9000 to prepare a CD for production. This represents a one-time fixed cost. Manufacturing, marketing, royalty payments, and so on are variable costs. Suppose the variable costs total $4.50 per CD. The CDs are sold for $6.75 each. How many CDs must be sold for the income from CD sales to equal the total costs.

6 Write an equation for the income (i), based on the number of CDs sold (N).

7 Write an equation for the total costs (C), based on the number of CDs produced (N).

8 Set the expression for i equal to the expression for C, and then isolate the variable N to find out what value of N is needed to make i equal to C.

9 You will receive a 5% commission on the sale of a condominium for a customer. The customer would like to obtain at least $60,000 from the sale. Use the formula below to determine the lowest selling price for the condominium that will allow the customer to receive the amount asked for. (Round your answer to the nearest $100.)

$$A = (100\% - c) \times s$$

A is the amount desired by the customer from the sale.
c is the commission that you will get from the sale.
s is the selling price.

The "Rule of 72" is a way to estimate the effect of compound interest on an investment. The rule states that you divide the annual percentage interest rate (expressed as a percent, not a decimal) into 72 to find the approximate number of years needed to double the value of your investment.

10 Write the "Rule of 72" as a formula. Be sure to identify the variables that you choose to use.

11 Use your formula to find out how many years it would take to double a $500 deposit in an account that paid an annual interest rate of 8.5%.

12 Use your formula to find out what interest rate would be needed to double an investment in five years.

You are told to "fax" (that is, electronically transmit a facsimile over the telephone lines) a document to your overseas branch office in Paris, France. The phone company charges $1.94 for the first minute. After the first minute, the charge is $1.09 per minute plus a 3% tax on the total charges. On the other hand, a 2-day postal delivery can be made for $22.

13 Write a formula that can be used to compute the total cost of a phone call of *n* minutes, when *n* is greater than one. Be sure to define your variables. What does a 10-minute phone call cost?

14 Rewrite the formula, isolating the total length of the call, *n.*

15 Use this last formula to determine how long a phone call could be to match the cost of the 2-day delivery of the document. How do you interpret this value for *n?*

You need to estimate the copying cost for a job that requires a large volume of copying. After a phone call to the copy shop, you find that each copy on white paper costs $0.045. On colored paper, it costs $0.065. In addition, there is a $5 handling fee for jobs that exceed 500 copies. Each packet of materials that you will duplicate has 5 colored pages and 27 white pages.

16 Write a formula you can use to determine the total cost for producing *n* packets, when *n* is greater than 500.

17 Suppose your supervisor tells you that this project has only $2000 allotted for copying costs. Use your formula to determine how many packets you can produce with this allotted amount.

18 Is there enough money in the budget to produce 1000 packets?

19 You are told that the demand for a certain product can be described by the equation

$$D = 4000 - 44C \text{ (for values of } C \text{ between 10 and 90).}$$

D is the demand, in pounds, for the product.
C is the cost per pound in cents.

In addition, the supply of this product from the supplier is described by the equation

$$S = -1340 + 134C \text{ (for values of } C \text{ between 10 and 90).}$$

S is the supply, in pounds, of the product.
C is the cost per pound in cents.

When the supply exceeds the demand, the price will go down. On the other hand, when the demand exceeds the supply, the price will go up. At what cost per pound will the price stabilize and the supply equal the demand?

HEALTH OCCUPATIONS

As children grow, dosages for medications gradually approach those for adults. Clark's rule is often used to determine the correct dosage for children.

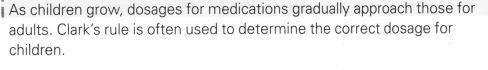

$$\text{Child's dose} = \text{Adult dose} \times \frac{\text{Weight of child in pounds}}{150 \text{ pounds}}$$

20 What dosage should be administered to a child weighing 28 pounds if the adult dose of a certain drug is 80 milligrams? Round your answer to the nearest whole milligram.

21 According to Clark's rule, how much would a child have to weigh to receive the adult dosage of a medication?

A person's vital capacity is the amount of air that can be exhaled after one deep breath. An estimate of the vital capacity for women can be obtained by the formula

$$V = 0.041h - 0.018A - 2.69$$

V is the vital capacity in liters.
h is the woman's height in centimeters.
A is the woman's age in years.

22 Determine an estimated vital capacity for yourself or a friend using the formula.

23 Suppose that by using a spirometer, the lung capacity of a healthy woman is measured at 3.4 liters. The woman is 5 feet 6 inches tall. Isolate the variable A in the formula. Use the given data to find a value for A and complete the sentence that follows:

This woman exhibited the breathing capacity of a __?__-year-old woman.

When a fair-skinned person comes into Marble's Tanning Salon, you normally advise him or her to start with 15 minutes' exposure, then gradually build up to a maximum of 23 minutes, increasing the exposure 1 minute every $2\frac{1}{2}$ days.

24 Which of the following equations would describe the exposure time, T (in minutes), for a fair-skinned client who has been tanning regularly for D days, prior to the point when he or she reaches the maximum time?

  a. $T = 23\,D - 15$
  b. $T = 2.5\,D^2 + 23$
  c. $T = 15\,D + 2.5$
  d. $T = 0.4\,D + 15$

25 How much exposure does the equation suggest you recommend on the 5th day of treatment? On the 10th day?

26 Solve the equation for D and use this new equation to determine how many days a client needs to reach the maximum 23 minutes of exposure.

The following table lists permissible sound exposures in the workplace established by the Occupational Safety and Health Administration (OSHA).

Hours of Exposure	Weighted per Day Sound Level (decibels)
8	90
6	92
4	95
3	97
2	100
1.5	102
1	105
0.5	110
0.25	115

A computer analysis of the data generates the equation

$$S = -2.817H + 108.9$$

S is the maximum permissible sound level in decibels.
H is the number of hours the person is in this environment.

27 Use the equation to compute the maximum possible sound levels for the hours of exposure listed in the table. Compare the computed sound levels to the values in the table.

28 Do you think the equation provides a good description of the table?

You are the manager of a drug testing laboratory. You want to compare the cost of breeding your own mice rather than purchasing them from a supply company. You can buy 6-week-old mice from the supply company for $4.90 each. Alternately, you estimate the cost of breeding the mice would be $830 each week in overhead and $0.60 per mouse each week for food. You need to find out how many mice would be needed each week to reach a point where the cost of purchasing equals the cost of breeding.

29 Write an expression for the cost of buying n mice from the supply company.

30 Write an expression for the cost of breeding n mice for one week.

31 You want to find how many mice it requires for the costs to be equal. You need to find the value of n that makes one expression equal the other. Set the two cost expressions equal to each other. Solve for the variable n. How many mice would the lab need per week for the decision to breed the mice to be cost-effective?

32 If more mice than this number are needed per week, would it be cheaper to purchase the mice or to breed them?

FAMILY AND CONSUMER SCIENCE

As an interior decorator, you frequently estimate costs for wallpapering a room. You always do the same set of calculations using different numbers for different rooms. For a given wall, you measure the height and the width. You divide the width of the wall by the width of the wallpaper to find how many strips will be needed. Then you multiply this number by the height of the wall to find the total length of paper needed.

33 Write a formula based on the above procedure that you could use to estimate the length of wallpaper needed to cover a wall. Be sure to define your variables.

34 If each roll contains 16 feet of wallpaper, modify your formula to provide an estimate of how many rolls of paper would be needed.

35 Try your formula for a wall that is 9 feet high and 14 feet wide, using wallpaper that is 27 inches wide. How many rolls would you estimate are needed to cover this wall?

36 What are some possible sources of error when using this formula?

Cosmetologists are frequently paid by the "salary and commission" method. The cosmetologist is paid a base salary plus a percentage of any money received that exceeds twice this base salary. The situation is modeled by the following equation:

$$G = S + C(i - 2S)$$

G is the gross pay (in dollars) for the period if i is greater than twice the salary amount.
S is the salary amount for the period.
C is the commission percentage rate (a decimal percent).
i is the gross income for services performed.

37 Suppose your weekly salary is $260, with a 15% commission on all sales over twice your salary. Use the formula to determine your gross pay if the gross income for your services totaled $685 during this week.

38 How much income for services is needed to have a gross pay of $300.

39 You examine the cost of bus travel for your vacation. After analyzing the fare schedule, you make a table of costs and mileages.

Mileage	One-way Fare
35	$ 5
100	10
145	17
180	23

Use a proportion to estimate the cost of a one-way fare for travel to a city that is 50 miles away.

Suppose you consider two job offers. One has a starting salary of $12,900 with raises of $1000 per year. The second offer is for a starting salary of $15,000 with raises of $650 per year.

40 Will these two salaries ever be the same? If so, when?

41 You hope to qualify for a promotion and a significant increase in salary after 5 years. Which of the two starting jobs would give you the higher salary at the end of five years?

INDUSTRIAL TECHNOLOGY

You are designing an exhaust fan for a small garage by using an electric motor connected to a large fan blade by a fan belt and two pulleys.

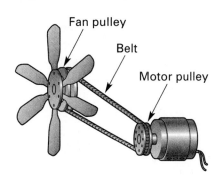

Fan pulley

Belt

Motor pulley

The approximate length of belt needed to connect the two pulleys is given by the formula

$$B = 2L + \frac{\pi}{2}(D + d)$$

B is the length of the belt.
L is the distance between the two pulley centers.
D is the diameter of the larger pulley.
d is the diameter of the smaller pulley.

42 The formula with definitions for the variables does not specify any units. You should assume the equation is true as long as the units from the variables are the same. If L, D, and d are measured in meters, what is the unit for B?

43 Compute the approximate length of belt needed to connect the two pulleys if the two pulley centers are 3 feet apart. The larger pulley is 8 inches in diameter, and the smaller pulley is 3.5 inches in diameter.

In close-up photography, the distance of an object from the lens determines how far the lens must be from the film. This requires special lenses and focusing mechanisms. These distances (all measured in the same units) are related by the formula

$$\frac{1}{f} = \frac{1}{p} + \frac{1}{q}$$

f is the focal length of the lens.
p is the distance of the object being viewed from the center of the lens.
q is the distance of the image formed on the film from the center of the lens.

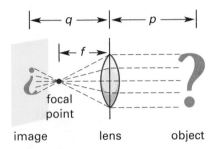

You have a 50-millimeter close-up lens. With this lens, the maximum distance you can position the lens from the film plate is 6.2 centimeters. You need to find the closest you can position an object to the lens to get a sharp photograph.

44 Rewrite the formula, isolating the variable representing the object distance.

45 Determine the object distance predicted by the equation for a lens with a focal length of 50 millimeters and a maximum image distance of 6.2 centimeters.

You need to make a wire-wound resistor with a resistance of 15 ohms using 22-gauge copper wire. From a table, you find that 22-gauge copper wire has a resistance of 0.0135 ohm per centimeter length. Thus, you can determine the resistance of a wire by multiplying the length of the wire by the number of ohms per centimeter.

46 Write a formula to determine the resistance of any length of 22-gauge copper wire. Be sure to define your variables and their units.

47 Use your formula to determine the resistance of 75 feet of 22-gauge copper wire.

48 Rewrite your formula, isolating the variable for length. Use this new formula to determine what length of 22-gauge copper wire you need to create the 15-ohm wire-wound resistor.

49 A particular electronics circuit lists several resistances R_2, R_3, and R_4 that are dependent on R_1. You cannot measure R_1 but you *can* measure R_2—it measures 110 ohms. Use this measurement to determine what the resistance for R_1 must be. Also determine what R_3 and R_4 must be.

Resistor	Resistance (ohms)
R_2	$1.5\,R_1$
R_3	$10\,R_1$
R_4	$0.25\,R_1$

Ohm's Law states that the voltage (in volts) is equal to the current (in amperes) times the resistance (in ohms).

50 Write Ohm's Law as an equation. Use V for the voltage, i for the current, and r for the resistance.

51 Suppose you observe various currents through a 3300-ohm resistor. Rewrite your equation for this condition.

52 Use the equation from Exercise 51 to determine what voltage you should measure if the current in the 3300-ohm resistor is 0.0175 ampere.

The National Fire Codes have guidelines for storing flammable liquids. An amount of 2 gallons per square foot is permitted in a storeroom with no fire protection (sprinkler or other type) and a rated fire resistance of 1 hour. The storeroom cannot be larger than 150 square feet.

53 Write an equation to express the relationship between the gallons of flammable liquid that can be stored and the area of the storeroom.

FLAMMABLE LIQUID

54 Identify the variables and the constants in your equation.

55 Use the equation to calculate the floor area required to store a maximum of 165 gallons.

56 How many gallons can you store in the largest storeroom?

You need to calibrate a variable, wire-wound resistor. You find that when the contact is at the minimum position, your resistor has a resistance of 4 ohms. Then, for every centimeter of travel away from this minimum position, the resistance increases by 2.5 ohms. This relationship continues up to a maximum of 29 ohms. At this resistance, the contact is at the farthest position—10 centimeters from the minimum.

57 Write a formula to compute the resistance of the resistor based on the distance from the minimum position.

58 How can you be sure of the validity of your formula?

59 Isolate the variable for the distance of the contact. Where should you place the contact to obtain a resistance of 22 ohms?

You are writing a contract for commercial sandblasting before repainting a building. To estimate the cost of the sandblasting, you determine the sand cost and the labor cost (to operate the sandblaster). You find that the sand costs $4 per bag and that it takes about 10 minutes to load and use each bag of sand. The labor cost for the operator is $40 per hour plus a one-time $110 equipment fee. Find an equation to estimate the total cost of the sandblasting, depending on the time.

60 Write an equation for the cost of the equipment used by an operator who works T hours.

61 Determine how much sand would be used in an hour (60 minutes) and hence how much it would cost per hour. Write an equation for the cost of sand when sandblasting for T hours.

62 Add the cost of the labor and equipment to the cost of the sand to find the total cost of sandblasting for T hours.

63 Use your equation to estimate the total cost if the job takes 2 hours. What is the cost for a 4-hour job and for a 6-hour job?

You work in a metal shop and must often drill equally spaced holes along a length of stock. The holes on the ends of the stock must be centered exactly 1 inch from the ends. The remaining holes will be equally spaced. A few examples are shown below.

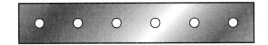

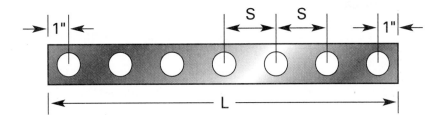

64 Write a formula for the spacings of the holes on a piece of stock that is L inches long and that must have a total of N equally spaced holes.

65 What spacing between holes (their centers) would be needed for 9 holes on a bar that is 10 feet long?

66 Suppose you use a $\frac{3}{4}$-inch drill bit (i.e., which has a $\frac{3}{4}$ inch diameter). Describe the layout of the holes on the bar if your equation yielded a spacing value of 0.75 inches. For nine holes in the bar, what length of stock would yield this value for the spacing?

Skills

Solve each equation. Check your answer.

1. $10a = -100$

2. $c + 25 = 16$

3. $3q + 5 = 19$

4. $-0.5m - 7 = 12$

5. $\dfrac{-20}{n} = \dfrac{12}{15}$

6. 15% of $x = 45$

7. 35% of $800 = y$

8. $\dfrac{1}{2} - r = -\dfrac{3}{4}$

9. $8.5 = 2.6 - 2d$

10. $5(3 + x) = 20$

11. $-9 = \dfrac{1}{2}(4 - y)$

12. $\dfrac{16 - x}{4} = -20$

13. $8y = 10 - 6y$

14. $-3(d + 6) = 2d$

15. $4(3 - d) = -(d + 5)$

Applications

Write and solve an equation for each situation.

16. A company pays one-half of the monthly premium on life insurance policies for its employees. If the company pays $140 a month for each employee, what is the total premium?

17. A TV and VCR together sell for $700. The price of the TV is $2\frac{1}{2}$ times the price of the VCR. What is the price of the VCR?

18. A maintenance department employs six workers. Each employee earns the same hourly rate. Last week each employee worked a total of 40 hours. The total payroll for the month was $2,640. What is the hourly pay for each worker?

19. One Saturday, Rhonda earned $10 more than Anne. If Rhonda and Anne earned a total of $124, how much did Anne earn?

20. One plumber charges $90 for a call and then $35 an hour. Another plumber charges $40 for a call and then $40 an hour. How many hours will the plumbers work if they make the same amount of money in the same amount of time?

Math Lab

21. Use the equation $C = \pi d$ and the measured values for C and d in the table to calculate an average value for π.

C	d
7.95 inches	2.53 inches
29.74 mm	9.47 mm
45.68 feet	14.54 feet

22. The perimeter of the irregular figure is 250 feet. What is the length of side x?

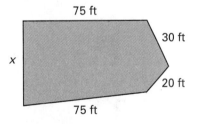

75 ft

30 ft

x

20 ft

75 ft

23. Use Algeblocks to solve the equation $4x - 2 = 10$

CHAPTER 5

WHY SHOULD I LEARN THIS?

The right solution can mean the difference between success and failure in a business. Successful problem solvers often use linear equations. Find out how you can use linear equations to increase your problem solving skills and make you more valuable to your employer.

GRAPHING LINEAR FUNCTIONS

OBJECTIVES

1. Graph data as points on a coordinate system.
2. Graph an equation.
3. Find the slope of a graphed line.
4. Find the intercepts of a graphed line.

In this chapter, you will draw the graphs of equations that represent straight lines. These equations are called **linear equations**.

A linear equation with two variables tells how those variables relate to each other. For example, the linear equation $d = 2r$ tells you that the diameter for any circle is twice the radius of that same circle.

$d = 2r$
Drawing

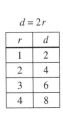

$d = 2r$

r	d
1	2
2	4
3	6
4	8

Table

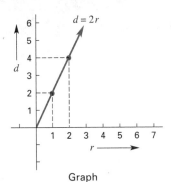

Graph

You can see this relationship if you *make a table* of values that satisfy the equation. Each pair of values in the table makes the equation true.

You can also see this if you *draw a graph.* The graph of a linear equation tells you how the two variables relate to each other, but in picture form. This graph is a line that shows how the radius and diameter are related.

As you watch the video for this chapter, notice how people on the job use equations to solve problems.

Coordinates on a Line

In Chapter 4, you used whole numbers, integers, and rational numbers to solve equations. A **rational number** is any number that can be written in the form $\frac{a}{b}$ where a and b are integers and b does not equal zero. *Terminating* decimals such as 0.5 are rational numbers because you can write them as fractions.

$$0.5 = \frac{5}{10} = \frac{1}{2}$$

Repeating decimals such as 0.333 . . . are also rational numbers. For example,

$$0.333\ldots = \frac{1}{3}$$

There are also decimals that do not terminate or repeat. Such decimals are called **irrational numbers**. The number π is an example of an irrational number.

$$\pi = 3.14159264\ldots$$
never terminates or repeats

Together, the rational and irrational numbers make up the set of **real numbers**. You use a number line to model the set of real numbers. Every point on the line is represented by exactly one real number, and every real number is represented by exactly one point.

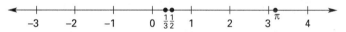

The numbers on the number line are referred to as the **coordinates** of the points. The points are **graphs** of the coordinates.

The graph of -2 is B.
The coordinate of C is 2.

Ongoing Assessment

What is the graph of 3? What is the coordinate of E?

For the rest of your study in Algebra this year, you will use the real number system. So, from now on, when the word *number* is used, it will mean *real number.*

Critical Thinking   One of the basic properties of the real number system is called **density**. That is, between any two real numbers there is always another real number. How can you show that the set of integers is not dense? How can you show that the set of rational numbers is dense?

Coordinates on a Plane

A plane is a flat surface that extends in all directions. The floor in your classroom is a model of a plane. You can also think of a piece of paper as a model for a plane.

From early civilizations, we know that people pictured pairs of numbers as points graphed on a plane. About 350 years ago, a French soldier, René Descartes, developed a system that used horizontal and vertical number lines to picture points and lines on a plane. His ideas worked so well that today we call this system the **Cartesian coordinate system** in his honor.

To draw a Cartesian coordinate system, begin by drawing a number line horizontally in the middle of your paper (if you want to use the whole sheet for the graph). This is the **x-axis**. Place the zero point about halfway across the page. This point is the **origin**. Label this point with the letter *O*.

Now draw another number line perpendicular (making a right angle) to the first, crossing the first line at the origin. This is the **y-axis**. Notice that the positive numbers are to the right of the origin for the horizontal line and up from the origin for the vertical line.

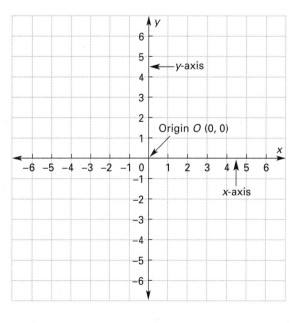

The size of the unit (the distance from zero to one) does not have to be the same on both number lines. But it is generally a good idea to draw the units the same unless you have a special reason for making them different. For example, you might need to have many small units on one line and only a few large units on the other.

The size of the unit is always the same on the positive and negative sides of a number line.

The horizontal number line and the vertical number line are both called **axes**. The horizontal axis is perpendicular to the vertical axis, and they cross at the origin, or **zero point**.

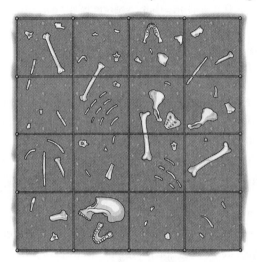

Pairs of numbers called **coordinates** represent every point on the Cartesian coordinate system. The first coordinate always tells how far to the right or left of the origin the point is located. The second coordinate in the pair always tells how far up or down from the origin the point is located. The number pair that names the origin or zero point is (0,0). Archaeologists use Cartesian coordinate systems to identify the location of artifacts.

EXAMPLE

Write the number pair that represents point A.

SOLUTION

To find the number pair that identifies point A, begin at the origin where the axes cross. Move to the right along the x-axis until you are under point A. As you move, count the number of units from the origin. Write down this number. Then move up along a line parallel to the y-axis until you reach point A. As you

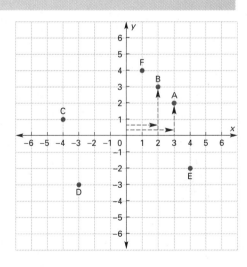

move, count the number of units from the *x*-axis. Write this number to the right of the first number. Place a comma between the pair of numbers and enclose them in parentheses. The pair (3,2) identifies point *A*.

Write the number pair that represents point *B*.

Did you write (2,3)? Notice that it *does* make a difference which number you write first. Point *A* is (3,2) and point *B* is (2,3). They are *not* the same point.

The pair of numbers that identifies a point in the Cartesian coordinate system is called an **ordered pair**. For example, the pair (3,2) that identifies point *A* is the ordered pair (3,2).

Ordered Pairs

The ordered pair of numbers that identifies any point is (x, y). The *x*-value of a point is the **x-coordinate** and is always written first. The *y*-value is the **y-coordinate** and is always written second.

ACTIVITY 1   Naming Points with Ordered Pairs

Write an ordered pair for each of the points *C*, *D*, *E*, and *F*. Remember, if the point lies to the left of the *y*-axis, its *x*-value is negative. If it lies below the *x*-axis, its *y*-value is negative.

Quadrants

The Cartesian coordinate system divides a plane into four quadrants.

Notice that the quadrants are numbered in a counter-clockwise direction.

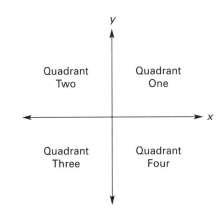

Use the ordered pairs you wrote for Activity 1 to help you choose the correct word (positive or negative) to complete these statements.

1 All the points in the first quadrant have x-values that are ___?___ and y-values that are ___?___.

2 All the points in the second quadrant have x-values that are ___?___ and y-values that are ___?___.

3 All the points in the third quadrant have x-values that are ___?___ and y-values that are ___?___.

4 All the points in the fourth quadrant have x-values that are ___?___ and y-values that are ___?___.

LESSON ASSESSMENT

Think and Discuss

1 Explain how the rational and irrational numbers are alike .

2 Explain the difference between a coordinate on a line and a coordinate on a plane.

3 How is a point on the Cartesian coordinate system named?

4 Why are the coordinates of a point written as an ordered pair?

Practice and Problem Solving

Write the ordered pair that represents each point.

5.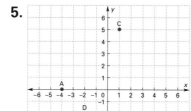

6.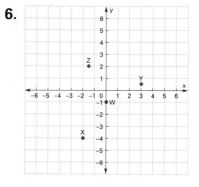

7. Identify the coordinates of each city.

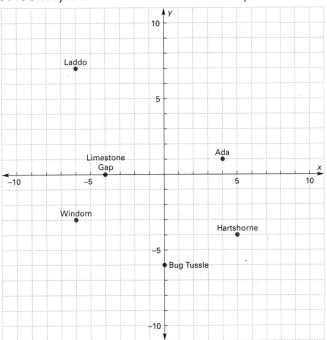

8. The data from a manufacturing process are shown on a graph. Identify the coordinates relating time deviation (*x*) to the length deviation (*y*).

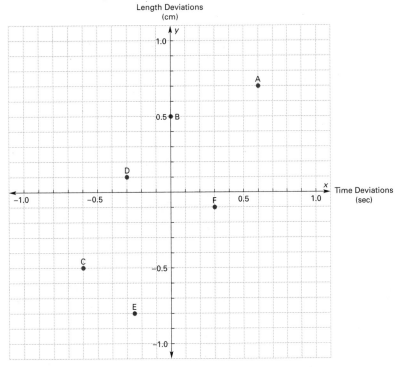

Give the quadrant where each point is graphed.

9. $(3, -5)$ **10.** $(-0.5, -1.3)$ **11.** $\left(-\dfrac{3}{4}, 4\right)$

Tell if the sentence is true or false. If it is false, give a counter-example. Let *a* and *b* represent integers.

12. If $a < b$, then $-a > -b$ **13.** $a + b = a - (-b)$

There are 10^2 centimeters in a meter. There are 10^3 meters in a kilometer.

14. How many centimeters are in a kilometer? Write your answer as a power of ten and in decimal notation.

15. How many kilometers are in a centimeter? Write your answer as a power of ten and in decimal notation.

The radius of a front wheel of a car is 42.3 centimeters. The radius of a rear wheel is 47.9 centimeters.

16. How far does the car travel for one turn of the rear wheel?

17. How far has the rear wheel traveled when the front wheel makes 500 turns?

The freshman class voted on the school color. The final vote was as follows:

Red 76   Blue 45   Green 90   Black 89

Write an equation and solve.

18. What percent voted for red?

19. How many votes would make 40% of the total vote?

20. What percent voted for black and blue combined?

21. What percent of the green vote is the blue vote?

Solve each equation for the given variable.
22. $6x - 33 = 15$ **23.** $4(y - 3) = -8$

LESSON 5.2 GRAPHING POINTS AND LINES

Ordered pairs of numbers identify all points located on the Cartesian coordinate plane. If you turn this idea around, you can say that every point on the Cartesian coordinate plane identifies an ordered pair of numbers.

The Cartesian Coordinate System

Every point on the Cartesian coordinate plane has a matching ordered pair of numbers. Conversely, every ordered pair of numbers has a matching point on the Cartesian coordinate plane.

There are many different coordinate systems. Any reference to a "coordinate system" in this book refers to the Cartesian coordinate system.

Graphing a Point

Draw a Cartesian coordinate system. Mark six equal units in all four directions. To locate the point $(-3, -3)$, begin at the origin. The x-value (-3) tells you to move three units to the left. The y-value (-3) tells you to then move three units down. Mark this as point G. You have located the point $G(-3, -3)$ as shown. To

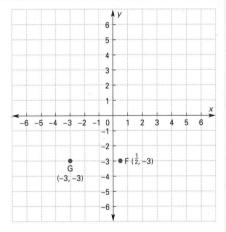

locate point F, with coordinates $(\frac{1}{2}, -3)$, move $\frac{1}{2}$ unit to the right along the x-axis. Now move 3 units down. The correct location of point $F(\frac{1}{2}, -3)$ is shown.

ACTIVITY 1   Locating Points

1   Draw a Cartesian coordinate system.

2   Label the horizontal axis the x-axis. Label the vertical axis the y-axis.

3 Locate and label the the ordered pairs: $A(1,5)$, $B(-1,5)$, $C(-1,-5)$, and $D(1,-5)$.

4 Draw the following line segments: *AB, BC, CD,* and *DA.*

5 Name the figure you have drawn.

Range and Scale

Choosing the range and scale units for the axes is an important step in making a graph. You can find the range needed for an axis by simply finding the difference between the smallest and largest values to be graphed along that axis. Once you know the range (and the size of the graph paper),

you then choose a scale unit so that the correct number of intervals fits along the axis.

Suppose you want to make a graph showing the distance traveled versus time for a race car moving at a constant speed of 200 miles per hour. You have the following data to graph:

Time	Distance
0.0 hr	0 mi
0.5 hr	100 mi
1.0 hr	200 mi
1.5 hr	300 mi
2.0 hr	400 mi
3.0 hr	600 mi
4.0 hr	800 mi

Also suppose you choose to plot time along the *x*-axis and distance along the *y*-axis. To establish the range for each axis, do the following:

- For *time along the* x-*axis*, find the difference between the largest and smallest values of the data for time. The range is 4 hours.

- For *distance along the y-axis,* find the difference between the largest and smallest values of the data for distance. The range is 800 miles.

To establish the scale unit, do the following:

- For the *scale unit along the x-axis* (the time axis), divide the range (4 hours) by the length of the x-axis. You should always allow margins of about 1.5 inches for labeling. For a standard page that is 8.5 inches wide, subtract about 3 inches for margins. This leaves about 5 inches of page space. Next, divide 4 hours by the 5 inches of page space. This gives a scale unit of 0.8 hour per inch, or about 1 hour per inch. Choose the scale unit of "one inch equals one hour." If the graph paper is marked off in small squares, you may want to count the nearest number of squares in an inch. Let the scale be, for example, "5 squares equal one hour."

- For the *scale unit along the y-axis* (the distance axis), divide the range (800 miles) by the y-axis length. For a standard page length of about 11 inches, you might choose 8 inches (leaving 3 inches for top and bottom margins) and get a scale of 100 miles per inch. Again, if you prefer to work with squares and there are about 2 squares to the inch, choose your scale unit as "1 square equals 50 miles."

With the range and scale chosen as outlined, the axes of your graph and your plotted points will look like those shown here. This graph is reduced in size to save space.

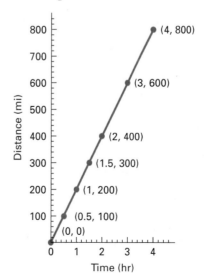

For the graph on the previous page, note the following:

- The scale units are large enough to give a good visual display of the graphed data.

- The scale units are large enough to let you determine distances traveled at the half-hour points.

The overall graph fits nicely on the 8.5-by-11-inch page. The graph provides ample room for the title and for the labels and scale units along each axis. Every graph should have a title, and scale units should always be labeled. That way, anyone who examines the graph will know what it represents.

ACTIVITY 2   Changing Scale

Use the same data given in the distance-versus-time table. Leave ample margin space. Use the given scale units to draw a graph on 8.5-by-11-inch paper.

1 Case 1.   *x*-axis: 1 inch = 4 hours
           y-axis: 1 inch = 400 miles

Can this be done? Why or why not?

2 Case 2.   *x*-axis: 1 inch = $\frac{1}{2}$ hour
           y-axis: 1 inch = 50 miles

Can this be done? Why or why not?

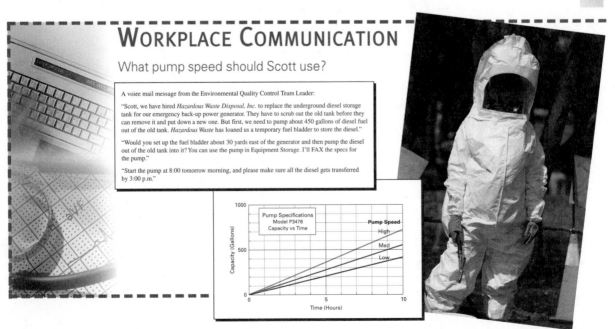

WORKPLACE COMMUNICATION

What pump speed should Scott use?

A voice mail message from the Environmental Quality Control Team Leader:

"Scott, we have hired *Hazardous Waste Disposal, Inc.* to replace the underground diesel storage tank for our emergency back-up power generator. They have to scrub out the old tank before they can remove it and put down a new one. But first, we need to pump about 450 gallons of diesel fuel out of the old tank. *Hazardous Waste* has loaned us a temporary fuel bladder to store the diesel."

"Would you set up the fuel bladder about 30 yards east of the generator and then pump the diesel out of the old tank into it? You can use the pump in Equipment Storage. I'll FAX the specs for the pump."

"Start the pump at 8:00 tomorrow morning, and please make sure all the diesel gets transferred by 3:00 p.m."

LESSON ASSESSMENT

Think and Discuss

1 Explain how to graph a point on the Cartesian coordinate plane.

2 Why is it important to understand how to determine the range and scale for the axes of a graph?

Practice and Problem Solving

Draw a Cartesian coordinate system. Locate and label each point.

3. $G(-3,4)$ **4.** $H(3,-4)$ **5.** $J(-3,-4)$ **6.** $K(3,4)$

7. Connect the points in order starting from quadrant one and ending in quadrant four. Describe the figure drawn.

Draw a Cartesian coordinate system. Locate and label each point.

8. $G\left(0,-3\frac{1}{2}\right)$ **9.** $H(-2,-2)$ **10.** $K(4,-2)$ **11.** $L\left(6,-3\frac{1}{2}\right)$

12. Connect the points in order starting from quadrant one and ending in quadrant four. Describe the figure drawn.

Draw a Cartesian coordinate system. Locate and label each point.

13. $S(0,0)$ **14.** $T(1,2)$ **15.** $U(3,6)$ **16.** $V(-2,-4)$

17. Connect the points. Describe the figure drawn.

18. Robots in a manufacturing plant move parts along an assembly line. Their movements are described in a coordinate plane. The range of Robot A is from $(7,8)$ to $(12,8)$. The range of robot B is from $(5,3)$ to $(5,-2)$. The range of Robot C is from $(7,-7)$ to $(12,7)$. Draw a Cartesian coordinate system. Plot the points for the range of each robot.

19. A jig-boring operation uses rectangular coordinates to establish each hole location. Draw a Cartesian coordinate system. Plot the centers of the holes specified by the ordered pairs $A(6,-2)$, $B(6,2)$, $C(10,1)$, and $D(10,-2)$.

Write each number in scientific notation.

20. 0.000385

21. $(1.89 \cdot 10^2)(7.4 \cdot 10^4)$

22. What is the value of $3m^3 - 2n$ when m is -3 and n is 12?

23. What is the value of $-2(r - t) + 5(t - r)$ when r is 2 and t is 4?

24. A carpenter is cutting circular blocks from plywood. If the diameter of the circle is 2.1 meters, what is the area of the plywood section being cut out?

25. If you borrow $1800 for 3 years at 6.5% simple interest, how much will you owe in interest?

Solve each equation for c.

26. $7c + 2 = -37$

27. $-2c - 8 = 12$

28. $\frac{2}{3}c + 11 = 5$

29. $\frac{c}{-4} = \frac{9}{3}$

30. 30% of $c = 120$

31. c% of $90 = 150$

LESSON 5.3 THE SLOPE OF A LINE

ACTIVITY 1   The y-axis

1 Graph these three points on the same Cartesian coordinate system.

$(0, -2)$, $(0,0)$, $(0,3)$

2 Describe the location of the three points.

All three points lie on the y-axis (the vertical axis). The x-value of each point is zero. However, the y-values are different. Any ordered pair that has an x-value equal to zero *must* identify a point somewhere along the y-axis. Thus, the equation $x = 0$ describes the y-axis.

ACTIVITY 2   The x-axis

1 Graph these three points on the same Cartesian coordinate system.

$(-2,0)$, $(0,0)$, $(3,0)$

2 Describe the location of the three points.

3 What is true about the y-value for each of these points? What is the equation that describes the x-axis?

ACTIVITY 3   Equal Coordinates

1 Graph these three points on the same Cartesian coordinate system.

$(-2, -2)$, $(0,0)$, $(3,3)$

2 Describe the location of the three points. Do all of these points lie on one straight line? What is true of the x-values and y-values of all these points?

3 Write an equation that describes this line.

You have used equations to describe three lines. You can also graph the lines on the same coordinate system.

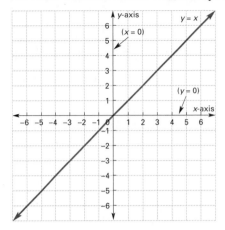

Slope

Imagine that each of the three lines ($x = 0$, $y = 0$, and $y = x$) is a hill. You have to climb each line. Which line is easiest to "walk up"? The "$y = 0$" line (or x-axis) is the easiest—it has no rise at all. You can say that the line $y = 0$ has a steepness of zero. This steepness is called the **slope** of the line. Thus, the line $y = 0$ has a slope of zero. Which line represents a hill so steep that it is impossible to climb? You cannot "walk up" the "$x = 0$" line (or y-axis) at all. The slope of this hill is so great that you cannot assign a number to it. The slope of the line $x = 0$ is undefined.

Which line represents a hill that is fairly steep, but one you could still climb? The "$y = x$" line has a slope somewhere between the x-axis (with a slope of zero) and the y-axis (with a slope that is undefined). How can you find the slope of the $y = x$ line?

The slope of a line is a measure of its steepness or "tilt." The steepness of a line (or a hill) is found by comparing its vertical rise to its horizontal run. A very steep road has a large amount of vertical rise for a given amount of horizontal run.

LESSON 5.3 THE SLOPE OF A LINE

ACTIVITY 1   The y-axis

1 Graph these three points on the same Cartesian coordinate system.

$(0,-2)$, $(0,0)$, $(0,3)$

2 Describe the location of the three points.

All three points lie on the y-axis (the vertical axis). The x-value of each point is zero. However, the y-values are different. Any ordered pair that has an x-value equal to zero *must* identify a point somewhere along the y-axis. Thus, the equation $x = 0$ describes the y-axis.

ACTIVITY 2   The x-axis

1 Graph these three points on the same Cartesian coordinate system.

$(-2,0)$, $(0,0)$, $(3,0)$

2 Describe the location of the three points.

3 What is true about the y-value for each of these points? What is the equation that describes the x-axis?

ACTIVITY 3   Equal Coordinates

1 Graph these three points on the same Cartesian coordinate system.

$(-2,-2)$, $(0,0)$, $(3,3)$

2 Describe the location of the three points. Do all of these points lie on one straight line? What is true of the x-values and y-values of all these points?

3 Write an equation that describes this line.

You have used equations to describe three lines. You can also graph the lines on the same coordinate system.

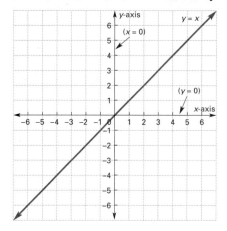

Slope

Imagine that each of the three lines ($x = 0$, $y = 0$, and $y = x$) is a hill. You have to climb each line. Which line is easiest to "walk up"? The "$y = 0$" line (or x-axis) is the easiest—it has no rise at all. You can say that the line $y = 0$ has a steepness of zero. This steepness is called the **slope** of the line. Thus, the line $y = 0$ has a slope of zero. Which line represents a hill so steep that it is impossible to climb? You cannot "walk up" the "$x = 0$" line (or y-axis) at all. The slope of this hill is so great that you cannot assign a number to it. The slope of the line $x = 0$ is undefined.

Which line represents a hill that is fairly steep, but one you could still climb? The "$y = x$" line has a slope somewhere between the x-axis (with a slope of zero) and the y-axis (with a slope that is undefined). How can you find the slope of the $y = x$ line?

The slope of a line is a measure of its steepness or "tilt." The steepness of a line (or a hill) is found by comparing its vertical rise to its horizontal run. A very steep road has a large amount of vertical rise for a given amount of horizontal run.

A road with a gentle slope has a small amount of vertical rise for the same amount of horizontal run.

Slope
The slope of a line is the ratio of the rise to the run.

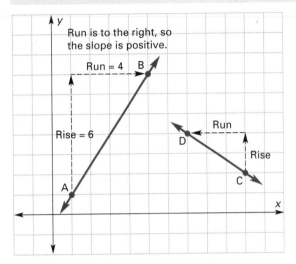

Lines have slopes that are positive, negative, or zero. To find the slope of a line, you need two points on the line. Imagine yourself walking from point A to point B. However, in this imaginary walk, you must first move up, and then go right or left; you cannot go diagonally.

Count the steps (or units) going up to find the rise. Then count the steps (or units) either left or right (the run) to reach B. If you move to the right, the number for the run is positive; if you move to the left, the number for the run is negative.

Once you know the value of the rise and run, write a fraction with the rise as the numerator and the run as the denominator. This fraction representing the ratio $\frac{rise}{run}$ is the slope of the line. The slope of line AB is

$$\frac{6}{4} \text{ or } \frac{3}{2}$$

Critical Thinking Why is the slope of the line passing through C and D negative?

EXAMPLE 1 Finding Slope

A surveyor places two stakes, A and B, on the side of a hill. Stake A is 20 feet lower than Stake B. If the horizontal distance between the stakes is 200 feet, what is the slope of the hill?

SOLUTION

Draw a diagram. The rise is 20 feet. The run is 200 feet. The slope is $\frac{20}{200}$ or $\frac{1}{10}$ or 0.1.

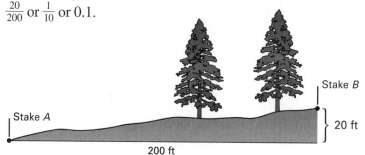

Stake B

Stake A

20 ft

200 ft

Look below at the graph of $y = x$. Pick two points, such as $P(1,1)$ and $Q(3,3)$. To move from P to Q, you can move up two units and then to the right two units. The slope is the ratio of the rise to the run.

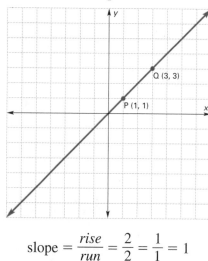

Q (3, 3)

P (1, 1)

$$\text{slope} = \frac{rise}{run} = \frac{2}{2} = \frac{1}{1} = 1$$

Thus, the slope of the line $y = x$ is 1.

The Slope Formula

The coordinates of any two points on a line determine its slope. The difference between the y-coordinates is the rise. The difference between the x-coordinates is the run. This gives a formula for finding slope.

Since the slope is the ratio $\frac{rise}{run}$, the slope can also be written in the following way:

$$\text{slope} = \frac{\text{difference of } y\text{-coordinates}}{\text{difference of } x\text{-coordinates}}$$

To find the slope of a line between two points, you can use the slope formula.

Slope Formula

If $A(x_1, y_1)$ and $B(x_2, y_2)$ are two points on line AB, then the slope of $AB = \frac{y_2 - y_1}{x_2 - x_1}$

When you use the slope formula to find the slope of a line between two points, be sure to subtract the coordinates in the same order.

EXAMPLE 2 The Slope Between Two Points

Find the slope of the line that joins $A(-2, -3)$ and $B(1, 4)$.

SOLUTION

Method 1   Make a sketch.

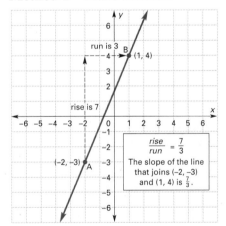

Start at the lower point, A. Move up to find a rise of 7 units. To reach point B, move to the right 3 units. This is a run of 3. Thus, the slope is $\frac{7}{3}$.

Method 2   Use the slope formula.

The difference in the y-coordinates is $4 - (-3) = 7$.

The difference in the x-coordinates is $1 - (-2) = 3$.

The slope is the ratio $\frac{7}{3}$.

LESSON ASSESSMENT

Think and Discuss

1 Describe the points that are located on the *y*-axis and on the *x*-axis.

2 Explain what the slope of a line means.

3 Explain how to find the slope of a line if you know its rise and its run.

4 Explain how to use the slope formula to find the slope of a line.

5 Explain why a slope can be positive or negative, and describe the line that models each slope.

Practice and Problem Solving

Find the slope of a line for the given rise and run.

6. rise 3, run -5 **7.** rise -2, run 4   **8.** rise 8, run 0

9. rise -10, run -2 **10.** rise 0, run 3   **11.** rise -2, run -10

A fencing contractor uses a scale drawing on a coordinate plane to calculate a bid on a job. Part of this calculation includes finding the slopes of the lines in a drawing. Find the slopes of the line segments with the given endpoints below.

12. $A(-3,4)$, $B(2,-1)$ **13.** $M(2,-5)$, $N(-1,-1)$

14. $C(2,3)$, $D(4,3)$ **15.** $X(0,0)$, $Y(2,4)$

16. $S(-6,2)$, $T(-6,5)$ **17.** $R(-8,-1)$, $S(3,-2)$

18. $L(-4,2)$, $M(3,-2)$ **19.** $P(7,10)$, $Q(-7,10)$

20. $G(-5,2)$, $H(5,-2)$

21. It is 220 miles from Johnson City to Putnam. The elevation of Johnson City is 3000 feet. The elevation of Putnam is 3500 feet. What is the average rate of increase in elevation per mile from Johnson City to Putnam?

22. What is the slope of the roof below?

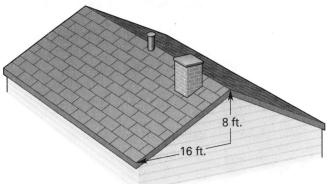

8 ft.

16 ft.

23. In a landing approach, an airplane maintains a constant rate of descent of 50 feet for every 500 feet traveled horizontally. What is the slope of the line that represents the landing approach of the plane?

Mixed Review

24. One light year is the distance light travels in one year. Light travels about 5.88×10^{12} miles in one year. Our galaxy is approximately 100,000 light years in diameter. What is the diameter of our galaxy? Write your answer in scientific notation.

For each situation, write and solve an equation.
25. The amount of water flowing over a dam at noon is 3.5 gallons per hour more than its rate at mid-morning. When the water flow was tested at noon, it had reached 48 gallons per hour. What was the rate of the water flow at mid-morning?

26. Keshia sells her inventory for twice what she pays. After expenses of $240 are deducted, Keshia finds she has $680 left. What did Keshia pay for her initial inventory?

27. Ramon is carpeting a rectangular room with a perimeter of 110 feet. One side of the room is 5 feet longer than the other. Find the length of the longer side.

Solve each equation. Check your answer.
28. $3d + (-4) = -1$ **29.** $x\%$ of $50 = 12$ **30.** $\dfrac{-5}{6} = \dfrac{10}{r}$

LESSON 5.4 GRAPHING LINEAR EQUATIONS

Comparing Slopes

The equation $y = 2x$ expresses a relationship between two variables x and y. In this relationship, the value of y *depends* on the value that is substituted for x. Thus, x is called the **independent** variable and y is called the **dependent** variable. If you select 1 as the value of the independent variable x, then 2 is the value of the dependent variable y. The equation $y = 2x$ is solved when you find the ordered pairs that make the equation a true statement. One solution of the equation is the ordered pair (1,2). Since the solution of $y = 2x$ is a set of ordered pairs, a table of values or a graph can represent $y = 2x$.

In a table, the first column is usually labeled as the independent variable. The second column is labeled as the dependent variable. Since you cannot list all the ordered pairs that make the equation a true statement, three to five pairs are usually enough.

x	y
1	2
0	0
-1	-2
-2	-4

After you complete the table, graph each pair of (x,y) values as a point. Draw the line that passes through all the points. Write the original equation next to or along the line.

Examine the figure that shows the graphs of $y = 2x$ and $y = x$ on the same pair of axes. Because the graphs are straight lines, $y = 2x$ and $y = x$ are called **linear equations**. Since the graph of the equation is determined by the table, you can find the slope of $y = 2x$ from the table.

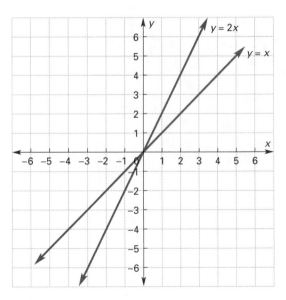

Critical Thinking Explain how to use the table to find the slope of the linear equation $y = 2x$.

ACTIVITY 1 The Slope of a Line

1 Use a table to graph each of the three equations on the same coordinate axes. Use four ordered pairs for each table.

   a. $y = x$ **b.** $y = 3x$ **c.** $y = 5x$

2 Use the slope formula to determine the slope of each line. Compare the slope of each line to the coefficient of the independent variable. What do you notice?

3 Guess the slope of $y = 6x$. Use a table to check your guess.

Notice that the graph of each equation is a straight line that passes through the point (0,0).

> ### Linear Equation
> The equation $y = mx$ is a linear equation. The graph of $y = mx$ is a line with slope m. The line passes through point (0,0).

Critical Thinking What happens to the slope of the line $y = mx$ as m increases?

EXAMPLE 1 Negative Slope

Compare the slope of $y = -2x$ with the slope of $y = 2x$.

SOLUTION

Make a table of values and graph $y = -2x$. First, choose four values for x. Then, use the equation to find the values for y.

x	y
-1	2
0	0
1	-2
2	-4

After you complete the table, graph each pair of (x, y) values as a point. Draw the line that passes through the points. Write the equation along the line.

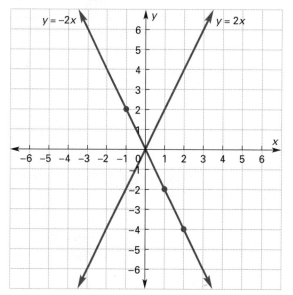

The steepness of these two lines is the same. However, they rise in opposite directions. What is the slope of each line? Notice that the slopes are opposite but equal values.

> **Positive and Negative Slope**
> A line that rises to the right has a positive slope. A line that rises to the left has a negative slope.

Look at your table of values for $y = -2x$. Notice that as x increases in value, y decreases. In this case, the graph of the equation has a negative slope.

Slope-intercept Form of an Equation

ACTIVITY 2   The y-intercept

1 Draw the graphs of these four equations on the same coordinate axes.

   a. $y = 2x$ **b.** $y = 2x + 1$ **c.** $y = 2x + 3$ **d.** $y = 2x - 2$

2 What is the slope of each line?

3 Where does each line cross the y-axis?

4 Where do you think the graph of $y = 2x + 5$ crosses the y-axis?

The y-value of the point where the line crosses the y-axis is called the **y-intercept**. The equation $y = 2x + 5$ is written in **slope-intercept form.** The slope is 2, and the y-intercept is 5.

> ### Slope-intercept Form of a Linear Equation
> $y = mx + b$ is a linear equation. The slope is m and the y-intercept is b. The line crosses the y-axis at the point $(0, b)$.

EXAMPLE 2 Graphing a Linear Equation

Graph the equation $2y = -3x + 6$ using slope-intercept form.

SOLUTION

First, write the equation in slope-intercept form. That is, solve the equation for y.

$$2y = -3x + 6 \qquad \text{Given}$$

$$\left(\frac{1}{2}\right)2y = \frac{1}{2}(-3x + 6) \qquad \text{Multiplication Property}$$

$$y = -\frac{3}{2}x + 3 \qquad \text{Distributive Property and Simplify}$$

The slope is $\dfrac{-3}{2}$, and the line crosses the y-axis at $(0,3)$.

Now draw the coordinate axes. Locate the point where the line will cross the y-axis. Since the slope is $\frac{-3}{2}$, you can find another point by moving down 3 and to the right 2. Locate that point. Connect the two points with a line.

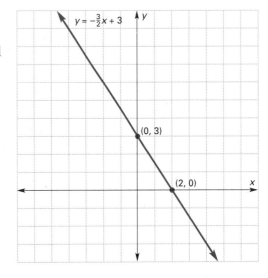

Write the equation $3x - y = -2$ in slope-intercept form. Graph the equation.

Linear equations model many problems. Once an equation is determined, you can use the graph to interpret the solutions.

Suppose a baker's recipe for cookies calls for a mixture of nuts and raisins and some chocolate chips. Some customers prefer lots of nuts; others would rather have more raisins. The recipe states that whichever way you mix them, when you add the number of cups of raisins to the number of cups of nuts, and then add this to $\frac{3}{4}$ cup of chocolate chips, you must have a total of 3 cups. How can you write an equation that models this situation? How would you draw the graph?

- First, translate the problem into sentence form:

   The number of cups of raisins plus the number of cups of nuts plus $\frac{3}{4}$ cup of chocolate chips must equal 3 cups.

- Second, let the letter r represent the number of cups of raisins and n represent the number of cups of nuts. Now you can change the sentence to the equation: $r + n + \frac{3}{4} = 3$.

To graph the equation, choose one of the variables as the independent variable. For example, let n be the independent variable. Then r takes the position of y, and n takes the position of x. To write the equation in slope-intercept form, isolate r on the left side of the equation.

$$r + n - n + \frac{3}{4} = 3 - n \qquad \text{Subtraction Property}$$

$$r + \frac{3}{4} - \frac{3}{4} = 3 - n - \frac{3}{4} \qquad \text{Subtraction Property}$$

$$r = -n + 2\frac{1}{4} \qquad \text{Simplify}$$

By comparing the equation $r = -n + 2\frac{1}{4}$ with the slope-intercept form $y = mx + b$, you can see that the slope (m) is -1. So you would expect this line to be higher on the left. What numbers do you want to use on the r and n axes? Since the values for both will be small positive numbers, including fractions, you might choose a fairly large (1 inch or so) distance for your unit. Draw the coordinate system on your paper, mark the units, and label the axes.

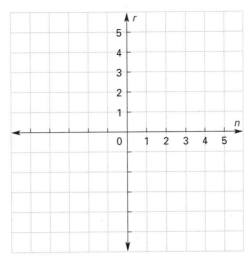

You can use the *y*-intercept to locate one point and use the slope to find another. Or you can make a table of several values. With only two ordered pairs, you can draw a straight line, but with several additional values, you can check your work. If all points are not on the same straight line, you have an error. If this happens, go back and check your arithmetic—and the way you made the graph. Graph your values and draw the line.

The graph of the equation $r = -n + 2\frac{1}{4}$ is a line that lies in quadrants one, two, and four. But the situation in the problem makes sense only in the first quadrant, where both r and n are positive numbers. As a result, the line you graph to fit the problem is limited to the first quadrant. Use a solid line in the first quadrant to emphasize where the data in the problem apply.

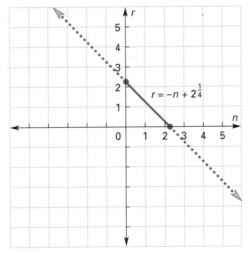

You can tell roughly what the temperature (T) is on a summer evening if you hear a cricket chirping and count how many times (N) it chirps in one minute. The formula to find the temperature in degrees Fahrenheit is the following:

$$T = \frac{1}{4}N + 40$$

1 Draw a graph of the formula. Use your graph to find the temperature if you count 100 chirps per minute. Find how many chirps you might hear if the temperature is 95 degrees.

2 What are reasonable values for T and N in this problem? Compare your graph to the one here.

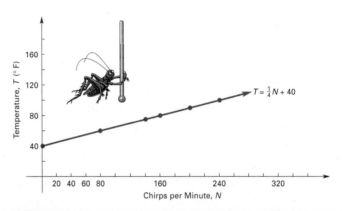

CULTURAL CONNECTION

Hypatia, the first woman mentioned in the history of mathematics, wrote about the work of a mathematician known as Diophantus of Alexandria. Sometime in the third century BCE, Diophantus wrote a text about arithmetic. In his text, Diophantus worked with equations that have more than one whole number solution. The equations are called Diophantine equations. Consider this problem.

In a pet shop there are several kittens and birds. The shopkeeper counted exactly 20 legs in the shop. How many kittens and birds does the shopkeeper have?

Let x represent the number of birds and y the number of kittens. The linear equation $2x + 4y = 20$ models this situation.

There are an infinite number of solutions to the equation. But the solution must make sense. How many kittens and birds can the shopkeeper have?

LESSON ASSESSMENT

1 How can you use an equation to make a table?

2 How can you use a table to graph an equation?

3 Explain why m is the slope of the equation $y = mx$.

4 Explain why the graph of the linear equation $y = mx + b$ crosses the y-axis at $(0, b)$

5 How can you graph the equation $3y + x = 9$ using slope-intercept form?

Practice and Problem Solving

Use a table to graph each equation. Give the slope and y-intercept of each graph.

6. $y = x + 2$ **7.** $y = -3x + 5$ **8.** $y = 2x - 4$

9. $x + y = 8$ **10.** $2x - y = 6$ **11.** $5x = 2y - 3$

12. $y = -5x - 3$ **13.** $y = \frac{1}{5}x - \frac{2}{3}$ **14.** $y - x = 4$

An asphalt company has found the equation $G = 10 + 20L$ approximates the number of trucks of gravel required for surfacing three-lane city streets. Let G represent the number of trucks of gravel required and L represent the length of the street in kilometers.

15. What are the slope and y-intercept of the equation?

16. Complete a table for the street lengths and the number of trucks of gravel. Use street lengths between 1 and 10 kilometers.

17. During one week, the asphalt company used 130 truckloads of gravel. How many kilometers of street did the company resurface?

In an experiment on plant growth, a certain species of plant is found to grow 0.05 centimeters per day. The plant measured two centimeters when the experiment started. Let H represent the ending measurement of the plant. Let d represent the number of days during which the experiment takes place.

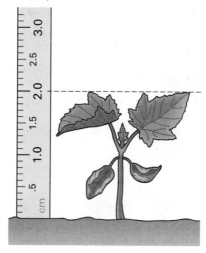

18. Write an equation that models the growth of the plant.

19. What are the slope and y-intercept for the equation?

20. At the end of the experiment, the plant was 3.3 centimeters tall. How many days did the experiment last?

Mixed Review

Write and solve an equation for each situation.

21. Paul is going to use 30% of his savings to make a down payment on a car. He can make a down payment of $2500. How much does Paul have in his savings account?

22. Tamara charges $6 per hour to clean windows. She also receives $3 for transportation to the job. One Saturday, Tamara earned $39. How many hours did Tamara work?

23. Nat has $50 and is saving money at $15 per week. Marti has $15 and is saving money at $20 per week. In how many weeks will they have the same amount of money?

24. At 7 AM, Jared notices that the temperature is 15°C. The weather forecaster just reported that from 5 AM to 7 AM, the temperature rose 5 degrees. What was the temperature at 5 AM?

LESSON 5.5 THE INTERCEPTS OF A LINE

Finding the Intercepts

The equation $y = mx + b$ is written in slope-intercept form. You know that the line representing this equation has a slope of m and crosses the y-axis at $(0,b)$. That is, the y-intercept is b. What is the y-intercept of $2x + y - 6 = 0$? To find the y-intercept, rewrite the equation in slope-intercept form.

$$2x + y - 6 = 0 \qquad \text{Given}$$

$$y = -2x + 6 \qquad \text{Why?}$$

The graph of $y = -2x + 6$ crosses the y-axis at $(0,6)$. Thus, the y-intercept is 6. What is the x-intercept of the graph?

The x-intercept is the point where the graph crosses the x-axis. All along the x-axis, the y-values are zero. Substitute zero for y and solve for x to find the x-intercept.

$$y = -2x + 6 \qquad \text{Given}$$

$$0 = -2x + 6 \qquad \text{Substitute zero for } y$$

$$2x = 6 \qquad \text{Addition Property}$$

$$x = 3 \qquad \text{Division Property}$$

The line crosses the x-axis at the point $(3,0)$. The x-intercept is 3.

Since the x- and y-intercepts represent two points on the line, you have another way to graph a linear equation. Graph the intercepts and connect them with a line.

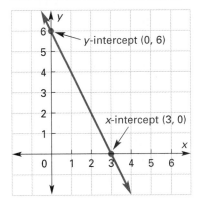

To check your graph, calculate one more point and graph it to be sure it is on the same line. For example, choose 1 for x. Then substitute in the equation $y = -2(1) + 6$. The result is the point $(1,4)$. Does this point lie on the same line determined by the intercepts?

It is easy to find the intercepts for the graph of a linear equation. To find the y-intercept, replace x by zero and solve for y. To find the x-intercept, replace y by zero and solve for x.

The equation $y = -2x + 6$ models this workplace situation. A tank is filled with 6 gallons of water. A pump begins to remove the water at a rate of 2 gallons per minute. How long will it take the pump to empty the tank? Here, the slope represents the pumping rate. What does the y-intercept represent? What does the x-intercept represent?

Ongoing Assessment

What are the intercepts for the equation $3x - 5y = 15$?

Change of Scale

Sometimes the slope and the intercepts of an equation make it necessary to change the scale of the graph.

EXAMPLE Changing Scale

Use the intercepts to graph $y = 20x + 25$.

SOLUTION

To find the y-intercept, replace x by zero. The result is a y-intercept of 25.

To find the x-intercept, replace y by zero. The result is the x-intercept $\frac{-25}{20}$ or $\frac{-5}{4}$.

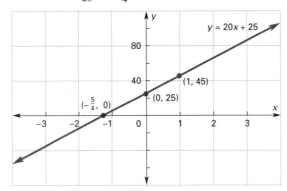

To check the graph, find one more point. If x is replaced by 1, y is equal to 45. Make sure the line passes through (1,45).

Since the slope of the line (20) is very steep, each axis has a different scale. This scale keeps the graph for $y = 20x + 25$ from being too close to the y-axis. By compressing the numbers along the y-axis (so that 40 units cover the same distance as one unit along the x-axis), the graphed line appears to have a very gentle rise.

Critical Thinking   Explain how changing the scale can make it hard to interpret the graphed data.

ACTIVITY   **Rate of Change**

The A-1 Cable Company installs underground cable. They charge $650 for the first 50 feet and $6 per foot for each additional foot. Let L represent the length of the cable installed (after the first 50 feet). Let d represent the number of dollars in the total cost.

1 Write an equation to describe the cost of installing cable.

2 Which variable is the

    a. independent variable?   **b.** dependent variable?

    Explain your answer.

3 Write the cost equation so that it is in slope-intercept form with the dependent variable (d) alone on one side of the equal sign.

4 Graph the equation.

5 What is the slope of the line? What is the y-intercept? What is the meaning of the slope and y-intercept in this situation?

6 Which part of your graph makes sense for the workplace situation?

7 Use your graph to read the approximate cost of installing a total of 175 feet of underground cable.

8 Check your answer. Show your check. Does your answer make sense?

LESSON ASSESSMENT

1 Explain how to find the *y*-intercept of a line.

2 Explain how to find the *x*-intercept of a line.

3 Give an example of a situation where the slope or intercepts would result in a change of scale of the graph.

Practice and Problem Solving

Find the *x*- and *y*-intercept for each line.

4. $y = 2x + 4$

5. $y = -3x + 5$

6. $y = 5x - 3$

7. $y = \frac{1}{2}x - 4$

8. $y = \frac{3}{4}x + \frac{1}{2}$

9. $y = 1.5x - 2.5$

10. $3x + y = -12$

11. $4x - y = -10$

12. $y - x = 5$

13. $2x - 3y = 12$

14. $5x + 3y = 15$

15. $x - 3y = -8$

Choose the independent and dependent variables. Then, write and graph an equation for each situation.

16. To convert Celsius to Fahrenheit degrees, multiply the Celsius temperature by 1.8 and add 32.

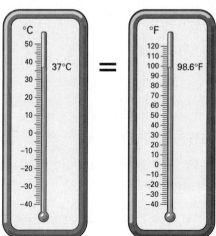

17. The distance from the start line after a certain amount of time if a runner is given a 2-mile head start and runs 4 miles per hour.

18. Lengths and widths of rectangular lawns with perimeters of 100 meters.

19. Amount earned if Anna mows lawns for $8 per hour and charges $6 for cleaning each yard.

20. Total cost if a seed company charges a fixed rate of $2.50 per pound plus a one-time $0.50 packaging charge.

Mixed Review

Write the answer as a power of ten.
21. $10^{-3} \cdot 10^{-2}$ **22.** $10^5 \cdot 10^{-5}$ **23.** $10^6 \div 10^{-2}$

24. Tim invested $5250 at 5.5% simple interest for 18 months. How much interest did Tim receive?

25. Draw a circle so that it fits into a square exactly. If the length of one side of the square is 625 centimeters, what is the area of the circle?

Evaluate each expression when x is -4 and y is -2.
26. $5x - 3y$ **27.** $x(y - 5) + 3$ **28.** $x^2 - y^3$

Catenya keeps a record of her earnings from the utility company. There are several ways Catenya can predict how much she will earn for any number of hours that she works. For example, she can write an equation or draw a graph.

Hours	Earnings
7	$52.50
4	$30.00

To draw the graph, plot the two points and draw a line through them. What is the scale? Why is the graph drawn only in the first quadrant?

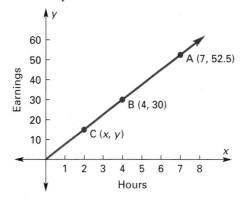

Using the Slope Formula

To write an equation, you will need to use the slope formula. The points $A(7, 52.5)$ and $B(4, 30)$ are on the line. Thus, the slope (m) of the line is

$$m = \frac{y_2 - y_1}{x_2 - x_1} = \frac{30 - 52.5}{4 - 7}$$

$$= \frac{-22.5}{-3}$$

$$= 7.5$$

You can represent every point on the line by an ordered pair (x, y). The slope of the line between $C(x, y)$ and $B(4, 30)$ is also 7.5 or $\frac{7.5}{1}$. Use the slope-intercept formula again. This time, use points C and B and solve for y.

$$\frac{7.5}{1} = \frac{y - 30}{x - 4} \qquad \text{Slope Formula}$$

$$7.5(x - 4) = y - 30 \qquad \text{Proportion Property}$$

$$7.5x - 30 = y - 30 \qquad \text{Distributive Property}$$

$$y = 7.5x \qquad \text{Addition Property}$$

The equation $y = 7.5x$ models Catenya's earnings.

In this equation, y is the dependent variable, and x is the independent variable. The equation shows that Catenya receives $7.50 for each hour she works. The amount of earnings (y) is a function of the time Catenya works (x).

The linear equation $y = 7.5x$ is a **linear function**. In this situation, the slope of the line represents a **rate of change** of $7.50 in the linear function. When x increases by 1, y increases by 7.5. This type of a rate of change is called **direct variation**.

EXAMPLE 1 Finding the Equation of a LIne

Find the equation of the line passing through $M(3,1)$ and $N(1,-3)$. Describe the graph connecting the points.

SOLUTION

Find the slope.
$$m = \frac{y_2 - y_1}{x_2 - x_1}$$

$$= \frac{-3 - 1}{1 - 3}$$

$$= 2$$

Let $P(x,y)$ be any point on the line. Use $M(3,1)$ as the second point.

$$2 = \frac{y - 1}{x - 3} \qquad \text{Given}$$

$$2(x - 3) = y - 1 \qquad \text{Proportion Property}$$

$$2x - 6 = y - 1 \qquad \text{Distributive Property}$$

$$y = 2x - 5 \qquad \text{Simplify}$$

The graph of the linear function $y = 2x - 5$ is a line with slope 2 and y-intercept -5.

> **Linear Function**
>
> An equation written in the form
>
> $$y = mx + b$$
>
> is a linear function. The rate of change of the function is m.

Ongoing Assessment

Find the equation of the line passing through $D(4,3)$ and $E(1,2)$ and describe its graph.

Horizontal and Vertical Lines

ACTIVITY 1   Horizontal Lines

1 Find the equation of the line passing through $(1,4)$ and $(2,4)$.

2 Find the equation of the line passing through $(-2,-3)$ and $(1,-3)$.

3 What do the slopes of the lines have in common?

4 Graph each equation on the same coordinate system.

5 What do the graphs have in common?

You can write each equation in Activity 1 in the form $y = b$. The equation can also be written as $y = mx + b$, where m is 0 and b is a constant. Thus, $y = b$ is a constant function. Its graph is a horizontal line, parallel to the x-axis. The value of b tells you how far above or below the origin the graph is drawn.

ACTIVITY 2   Vertical Lines

1 Find the equation of the line passing through $(4,1)$ and $(4,2)$.

2 Find the equation of the line passing through $(-2,-2)$ and $(-2,1)$.

3 What do the slopes of the lines have in common?

4 Graph both equations on the same coordinate system.

5 What do the graphs have in common?

You can write each equation in Activity 2 in the form $x = c$. However, the equation $x = c$ is not a function. In the next chapter you will find out why. The graph of $x = c$ is a vertical line, parallel to the y-axis. The value of c tells you how far to the right or left of the origin the graph is drawn.

Critical Thinking   Compare these graphs to the equations in the Activities. How are they alike or different?

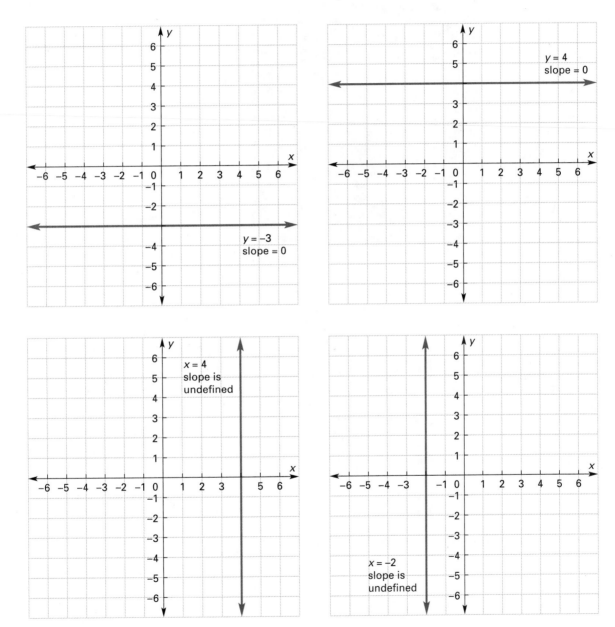

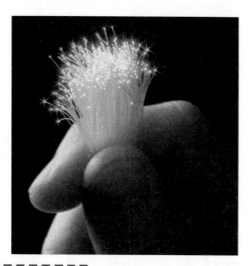
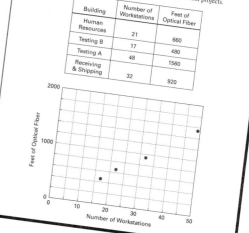
LESSON ASSESSMENT

Think and Discuss

1 Explain how to find the equation of a line passing through two points if you know the coordinates of each point.

2 What is a linear function?

3 How are the rate of change of a function and the slope of a line related?

4 Explain why a horizontal line represents a linear function.

Practice and Problem Solving

Find the equation of the line passing through the given points.

5. (4,5), (2,3)

6. (3,−2), (6,0)

7. (−1,8), (−3,5)

8. (−2,4), (4,4)

9. (−7,−2), (−6,−1)

10. (−9,6), (−9,−6)

A moving van travels 325 miles in 5 hours. The next day, the same van travels 520 miles in 8 hours. Let x represent the time and y represent the distance traveled.

11. Write the two ordered pairs from the times and distances given. Find the rate of change in the distance the car travels.

12. Write a linear function that models the situation.

Todd is selling tickets for the school play. It costs $5 for an adult ticket and $3 for a child's ticket. Todd needs to sell a total of $65 in tickets. Let x represent the number of adult tickets sold. Let y represent the number of child's tickets sold.

13. Write a linear function that models the number of adult and child's tickets Todd must sell to reach his goal.

14. Graph the linear function.

15. Suppose (10,5) is a point on your graph. What does the ordered pair represent in the problem?

Hook's Law states that the distance (d) a spring is stretched is directly proportional to the force (F) applied. The formula is often written in the form $F = kd,$ where k is called the spring constant. Suppose a force of 500 pounds stretches a truck's suspension spring one inch.

16. What is the spring constant of the truck spring?

17. Write a linear function that models the spring displacement as a function of applied force.

18. How far will the spring stretch under a force of 700 pounds?

19. How much force is needed to stretch the spring 3 inches?

Represent each situation with a positive or negative number.

20. A profit of $1250

21. A weight loss of 12 pounds

22. A salary increase of $2/hr

23. A temperature fall of 9 degrees

Solve each equation. Check your answer.

24. $5t - 12 = -t$ **25.** 45% of $m = 90$ **26.** $\dfrac{-20}{d} = \dfrac{7.5}{15}$

Write and solve an equation for each situation.

27. Winston finished 20% of his job assignment the first day. In that time, he was able to paint 24 outside window casings. How many window casings will Winston paint at the end of his assignment?

28. A draftsman is completing a design that uses isosceles triangles. In one of the triangles, the measure of the vertex angle is represented by a. The measure of each base angle is represented by the expression $2a + 20$. Find the size of each angle in the triangle.

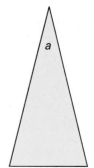

LESSON 5.7 USING A GRAPHICS CALCULATOR

Most graphics calculators graph linear functions written in the form $y = mx + b$. Check out the graphics calculator you will be using. Although the instructions given in this text might differ somewhat from how your calculator works, the ideas should be similar. Here is one way to graph $y = 3x + 1$.

Press the $\boxed{Y=}$ key. Then enter $3x + 1$. Now press the $\boxed{\text{GRAPH}}$ key. You should see the graph of $y = 3x + 1$ appear on the screen.

The graphics calculator window is very small. Thus it is important that a "friendly window" (one with a readable scale) is selected before you graph a function. The $\boxed{\text{WINDOW}}$ or $\boxed{\text{RANGE}}$ key is used to set the maximum (max) and minimum (min) range and the scale (scl) division. The standard setting for this text will be

$X\text{min} = -10$ $X\text{max} = 10$ $X\text{scl} = 1$

$Y\text{min} = -10$ $Y\text{max} = 10$ $Y\text{scl} = 1$

Ongoing Assessment

Use your graphics calculator to graph $y - 2x = -3$.

Critical Thinking Explain how to use your graphics calculator to graph $y = 2x + 2$ and $y = -5x + 3$ on the same coordinate plane.

The $\boxed{\text{TRACE}}$ key is used to find the coordinates of any point on the graph of the linear function.

ACTIVITY 1 Graphing a Linear Function

1 Graph the function $y = 3x + 1$ on your graphics calculator.

2 Press the $\boxed{\text{TRACE}}$ key.

3 What is the result?

4 Press the left and right arrow keys.

5 What is the result?

There is another important key on your graphics calculator.

1 Press the ⌑ZOOM⌑ key.

2 What is the result?

3 Explain how the ⌑ZOOM⌑ key results in a more accurate reading of the coordinates of a point.

LESSON ASSESSMENT

Think and Discuss

1 What are the advantages of using a graphics calculator to graph a linear function? What are the disadvantages?

Explain how to graph each equation on your graphics calculator.

2 $y = 3x - 1$

3 $2x + y = 6$

4 $y = \frac{1}{2}x + 5$

5 Reread the situation presented at the beginning of Lesson 5.6. Explain how to change the range or scale settings to get a friendly window for this situation.

Practice and Problem Solving

Use your graphics calculator.
Graph both equations on the same coordinate axis.

    a. $y_1 = 3x + 2$ **b.** $y_2 = 3x - 1$

6. What is the relationship between the lines?

7. What is the relationship between the slopes?

8. Write another equation with the same relationship. Graph the equations on the same coordinate axis to check your result.

Graph both equations on the same coordinate axis.

    a. $y_1 = -2x + 2$ **b.** $y_2 = \frac{1}{2}x - 1$

9. What is the relationship of the lines?

10. What is the relationship of the slopes?

11. Write the equation of a line perpendicular to the graph of $y = 4x - 3$. Graph the equation to check your result.

Mixed Review

Solve each equation.

12. $5a - 12 = -27$

13. $\dfrac{-6}{9} = \dfrac{r}{-12}$

14. 25% of $d = 80$

15. $\dfrac{2}{3}m + 7 = -15$

16. $-24 = 18 - q$

17. $5s + 4 = 13 - 4s$

Write each equation in slope-intercept form.

18. $5x + 2y = 20$ **19.** $3x - y = 2$ **20.** $2x = 9 + 3y$

Write and solve an equation for each situation.

21. Latisha invests $2500 at 6.5% simple interest. How much does she have at the end of 18 months?

22. Shareef sells model cars for $8. At one craft show, Shareef earned $1032. How many model cars did he sell?

23. The combined total company sales for this year and last year are $3,500,000. This year, sales are 1.5 times last year's sales. What are the sales for this year?

24. Mr. Williams is putting a fence around a rectangular garden. The perimeter of the garden is 300 feet. The length of the garden is twice the width. How long is the garden?

MATH LAB

Activity 1: Height and Volume of a Cylinder

Equipment   Calculator
1 lb coffee can
Graduated cylinder, 500 ml capacity
Vernier caliper
Ruler

Problem Statement

You will examine the relationship between the volume (V) and height (h) of a cylindrical can and plot data of volume vs. height.

The formula for the volume of a cylinder is $V = \pi r^2 h$.

Procedure

a Measure the inside diameter of the 1 pound coffee can. Calculate and record the radius in centimeters. To convert from inches to centimeters, use 2.54 centimeters equal to 1 inch. Calculate and record the area of the base of the coffee can.

b Measure 100 ml of water in the 500 ml graduated cylinder and pour it into the coffee can. Be sure the can is level. To check if it is level, measure the height of the water in the can at three different positions around the inside edge of the can. If all three measurements are equal, the can is level. Measure and record the height in centimeters of the water in the can. Record the volume of the added water.

c Repeat Step **b** by adding 100 ml of water four more times for a total of 500 ml of water. Be sure to record the height of the water and the total volume of water in the can after each addition of water. When you are finished, you should have five pairs of values for V and h in your data table.

d Graph your data. Graph the volume on the vertical axis and the height on the horizontal axis. Be prepared to defend your answers to the questions on the opposite page.

1. Is the relationship between volume and height linear?

2. Is the equation $V = \pi r^2 h$ linear if r is constant?

3. If the equation is linear, what is the slope of the line represented by the equation?

4. How does this slope compare to the calculated value from Step **a**?

5. Does the graphed line pass through the origin if you extend it? Should it?

6. What does the point on the line at the origin mean?

7. How would the slope change if you used a larger diameter coffee can?

Activity 2: Measuring in Inches and Centimeters

Equipment Calculator
Tape measure marked in inches and centimeters

Problem Statement

The relationship that converts measurements in inches (i) to centimeters (C) is

$$C = 2.54i$$

You will measure several lengths in inches and centimeters and plot corresponding pairs of measurements on a graph. You will interpret the graph to verify the value of the coefficient 2.54.

Procedure

a Measure and record the width of a sheet of paper in inches and in centimeters.

b Measure and record the length of a sheet of paper in inches and in centimeters.

c Measure and record the width of the classroom door in inches and in centimeters.

d Measure and record the height of the classroom door in inches and in centimeters.

e Measure and record the width of the teacher's desk in inches and in centimeters.

f Measure and record the length of the teacher's desk in inches and in centimeters.

g Graph your data. Graph the measurement in centimeters on the vertical axis. Graph the measurement in inches on the horizontal axis. *Caution:* Study the range of the data. Then choose the scales for the *x* and *y* axes. Make certain all the data will fit on the graph.

h Draw an unbroken line that best connects the six points on your graph. Is the graph a straight line? If a point is not on the straight line, double-check your measurements and graph for that point.

i Choose any two points on the graphed line—such as *A* and *B* in the drawing shown here. These points need not include the points you plotted to draw the graph. Based on the values of these points, subtract the smaller centimeter value from the larger centimeter value. The result is the difference in centimeter values for the two points. Label this on the graph as Δcm. [Note: Δ is the Greek letter delta. It is often used to indicate a *difference* in values.]

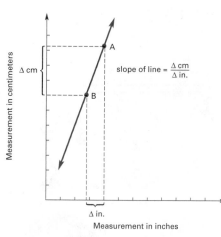

j For the same two points, and in the same order, find the difference in inch values. Label this on the graph as Δin.

k Divide the Δcm value by the Δin. value. This is the slope of the graphed line and is the value of *m* in the slope-intercept form of a linear equation $y = mx + b$. Compare your calculated slope to the value 2.54 in the equation 1 (cm) = 2.54 · 1 (in.).

l For your graphed line, what is the *y*-intercept? Does this value make sense?

Equipment   Calculator
1000 ml beaker
Graduated cylinder, 500 ml capacity
Spring scale with 500 gm capacity
String

Problem Statement

If the temperature is constant, the relationship between the weight and volume of liquid is a linear function. You will study this relationship by weighing different volumes of water and then calculating the slope and *y*-intercept.

Procedure

a   Tie a string around the top of the graduated cylinder. Weigh the graduated cylinder using the spring scale. Record this weight.

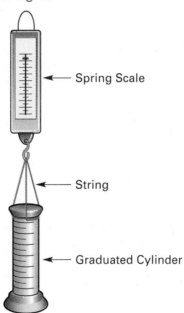

Spring Scale

String

Graduated Cylinder

b   Fill the 1000 ml beaker about half full of water.

c   Pour about 100 ml of water from the beaker into the graduated cylinder and weigh the water and graduated cylinder with the spring scale. Record the volume of water and the weight.

d Repeat Step **c** four more times until about 500 ml (but no more) is added to the graduated cylinder.

e Graph your data. Graph the weight on the vertical axis and the volume of water in the graduated cylinder on the horizontal axis.

f Draw an unbroken line connecting the points on your graph. Is the graph a straight line? If any point is not on the straight line, check your data and graph to be sure the point is correctly plotted.

g Choose two points next to each other. Subtract the smaller weight from the larger weight. This is the difference in weights. Label this difference as Δw. (Refer to Activity 2 for a drawing that shows a similar graph with similar measurements.)

h For the same two points, subtract the smaller volume from the larger volume. This is the difference in volumes. Label this difference as ΔV.

i Divide the difference in weight, Δw, by the difference in volume, ΔV. This is the slope of the graphed line and is the value of m in the slope-intercept form of a linear equation $y = mx + b$.

j What is the weight on the graph when the volume of water is zero? Compare this value to the weight of the empty graduated cylinder. This value is the y-intercept of the graph and corresponds to the b value in the slope-intercept form of a linear equation $y = mx + b$.

k Write a linear equation in slope-intercept form that shows the relationship between the weight and volume of given amounts of water as established by your data. Your equation should include the value for the constant m and the constant b.

MATH APPLICATIONS

The applications that follow are like the ones you will encounter in many workplaces. Use the mathematics you have learned in this chapter to solve the problems. Wherever possible, use your calculator to solve the problems that require numerical answers.

At a constant temperature, the weight of water increases as the volume increases. The graph below shows the relationship between water's weight and volume at 60°F.

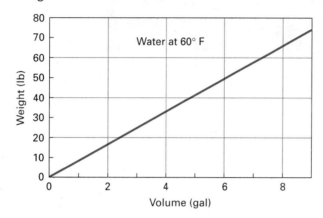

1 What is the weight of 2 gallons of water at 60°F? What is the weight of 6 gallons of water at 60°F?

2 What is the slope of the graphed line? What is the *y*-intercept of the graphed line?

3 Write the equation of the graphed line in slope-intercept form.

For each range of data below, use the maximum length of the axis given for a graph to determine a scale unit for that axis. For example, for a data range of 0 to 4 hours to fit an axis 4 inches in length, you can use a scale unit of 1 hour per inch.

	Data Range	Maximum Length of Axis	Scale Unit
4	0 to 800 miles	8 inches	
5	0 to 500,000 persons	5 inches	
6	0 to 60 seconds	12 centimeters	
7	25 to 30 minutes	10 centimeters	
8	80°C to 120°C	10 centimeters	

Pressure changes with depth in water and with altitude in air. The following tables list values of pressures at different depths below and altitudes above sea level.

| Water Pressure vs Depths | | Air Pressure vs Altitude | |
Depth (ft)	Pressure (Atm)	Altitude (1000 ft)	Pressure (Atm)
0	1	0	1.00
33	2	10	0.69
66	3	20	0.46
100	4	30	0.30
133	5	40	0.19
166	6	50	0.11
200	7	60	0.07
300	10		
400	13		
500	16		

9 Plot graphs of the two sets of data.

10 Does either graph show a linear relationship? If so, which graph(s)?

11 If any of the graphs are linear, identify the slope of the graph and the y-intercept of the graph.

12 Use your values of the slope and the intercept to write an equation for the linear relationship(s).

AGRICULTURE & AGRIBUSINESS

Steers weighing 800 to 900 pounds consume an average of 22.3 pounds of feed per day. These steers should gain weight at an average rate of 2.7 pounds per day.

13 Write an equation in slope-intercept form showing the number of days for a steer in this weight range to gain *W* pounds of body weight.

14 Write another equation in slope-intercept form to express the number of days needed for an average steer in this weight range to consume *F* pounds of feed.

15 Both equations from Exercises 13 and 14 define the number of days. Thus, you can set them equal to each other to obtain a new equation. Do this and obtain an equation that relates the pounds of feed consumed (*F*) to the pounds of weight gained (*W*). Isolate the variable for the pounds of feed consumed.

16 Use your equation from Exercise 15 to calculate the feed consumption needed for a 50-pound weight gain.

A plant's water requirement is defined as the pounds of water that must pass from the roots and out of the leaves to produce one pound of dry plant matter. The water requirement for corn is 368 pounds of water per pound of dry plant matter produced.

17 Write an equation in slope-intercept form that expresses the relationship between the yield of plant matter and the water requirement for corn.

18 Use your equation to calculate the water required to produce 8,000 pounds of dry corn plant matter.

19 An acre-inch of water is equal to 226,512 pounds of water. Use your equation to calculate the yield of dry corn plant matter that consumes one acre-inch of water.

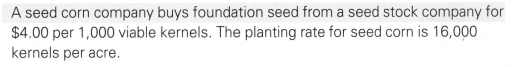

A seed corn company buys foundation seed from a seed stock company for $4.00 per 1,000 viable kernels. The planting rate for seed corn is 16,000 kernels per acre.

20 Write an equation in slope-intercept form for the relationship between the acres planted and the number of kernels of seed corn.

21 Write an equation in slope-intercept form that shows the relationship between the acres planted and the cost of the seed corn.

22 Use your equation from Exercise 21 to calculate the cost of the seed corn used to plant 120 acres.

A certain farming area has found that the cost per acre of loading hay to be closely related to the total labor hours required to complete the job. The following graph shows a plot of cost per acre versus labor hours per acre.

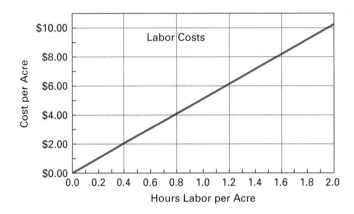

23 What is the slope of the graph? What rate of change does this slope represent?

Farmers can use various methods to load hay, each requiring different labor hours per acre. The labor required for three methods is as follows:

Trailed wagon and chute	1.54 hours per acre
Bale thrower	1.16 hours per acre
From ground by hand	1.97 hours per acre

24 What is the cost for loading 100 acres of hay from the ground by hand?

25 What are the costs for loading hay with a bale thrower and with a wagon and chute for 100 acres?

The various food supplies in a zoo are used at different rates. Suppose the current quantity and usage for three different types of foods are as listed below.

Food	Current Quantity (in pounds)	Usage per Day (in pounds)
Type A	700	40
Type B	640	60
Type C	480	30

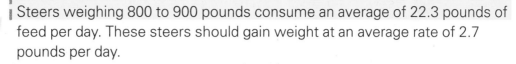

AGRICULTURE & AGRIBUSINESS

Steers weighing 800 to 900 pounds consume an average of 22.3 pounds of feed per day. These steers should gain weight at an average rate of 2.7 pounds per day.

13 Write an equation in slope-intercept form showing the number of days for a steer in this weight range to gain *W* pounds of body weight.

14 Write another equation in slope-intercept form to express the number of days needed for an average steer in this weight range to consume *F* pounds of feed.

15 Both equations from Exercises 13 and 14 define the number of days. Thus, you can set them equal to each other to obtain a new equation. Do this and obtain an equation that relates the pounds of feed consumed (*F*) to the pounds of weight gained (*W*). Isolate the variable for the pounds of feed consumed.

16 Use your equation from Exercise 15 to calculate the feed consumption needed for a 50-pound weight gain.

A plant's water requirement is defined as the pounds of water that must pass from the roots and out of the leaves to produce one pound of dry plant matter. The water requirement for corn is 368 pounds of water per pound of dry plant matter produced.

17 Write an equation in slope-intercept form that expresses the relationship between the yield of plant matter and the water requirement for corn.

18 Use your equation to calculate the water required to produce 8,000 pounds of dry corn plant matter.

19 An acre-inch of water is equal to 226,512 pounds of water. Use your equation to calculate the yield of dry corn plant matter that consumes one acre-inch of water.

A seed corn company buys foundation seed from a seed stock company for $4.00 per 1,000 viable kernels. The planting rate for seed corn is 16,000 kernels per acre.

20 Write an equation in slope-intercept form for the relationship between the acres planted and the number of kernels of seed corn.

21 Write an equation in slope-intercept form that shows the relationship between the acres planted and the cost of the seed corn.

22 Use your equation from Exercise 21 to calculate the cost of the seed corn used to plant 120 acres.

A certain farming area has found that the cost per acre of loading hay to be closely related to the total labor hours required to complete the job. The following graph shows a plot of cost per acre versus labor hours per acre.

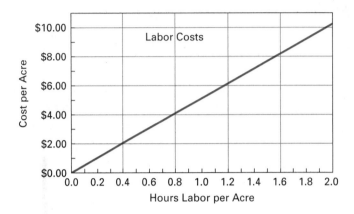

23 What is the slope of the graph? What rate of change does this slope represent?

Farmers can use various methods to load hay, each requiring different labor hours per acre. The labor required for three methods is as follows:

Trailed wagon and chute   1.54 hours per acre
Bale thrower   1.16 hours per acre
From ground by hand   1.97 hours per acre

24 What is the cost for loading 100 acres of hay from the ground by hand?

25 What are the costs for loading hay with a bale thrower and with a wagon and chute for 100 acres?

The various food supplies in a zoo are used at different rates. Suppose the current quantity and usage for three different types of foods are as listed below.

Food	Current Quantity (in pounds)	Usage per Day (in pounds)
Type A	700	40
Type B	640	60
Type C	480	30

26 Let x represent the number of days. Let y represent the number of pounds of food. Let the rate *usage per day* be the slope and *current quantity* be the y-intercept. Write linear equations for the food types that represent the number of pounds of food remaining after x days. (Put the equations in slope-intercept form.)

27 Choose a scale and draw a graph of the three lines. Label each line according to the food type. The slopes are negative. Why?

28 Use your graph to determine which of the three types of food will be the first to be consumed. How many days in the future will this occur? (Hint: This is the x-intercept for one of your graphed lines.)

A farmer can estimate the cost of transporting crops to market as a sum of the cost of fuel and the cost of the vehicle, as they relate to the distance to market. Fuel costs for the vehicle are about $0.095 per mile. The cost of the vehicle is a fixed value of $2500 per year.

29 Write an equation for the annual transportation cost as a function of total annual miles traveled.

30 What is the slope of your equation? What does this rate represent?

31 What is the meaning of the y-intercept of the equation?

32 Draw a graph of the equation.

BUSINESS & MARKETING

A researcher is analyzing past trends in consumer spending. The graph shown here displays some changes in consumer spending that took place over a period of ten years.

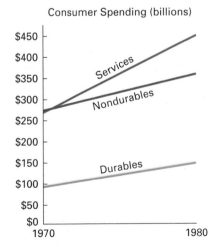

Consumer Spending (billions)

33 The line representing spending for durables and the line for nondurables appear similar. What feature of these two lines is nearly the same?

34 The two lines for services and nondurables are similar but have one feature that distinguishes them. What feature of these two lines is notably different?

A department store pays each of its senior sales clerks a weekly salary of $150 plus a 3% commission on the clerk's gross sales. This graph shows the pay scale.

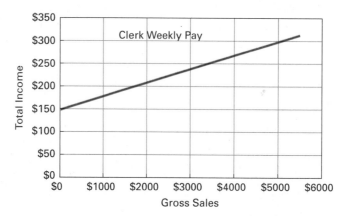

35 What is the y-intercept of the graph? What quantity does this intercept represent?

36 What is the slope of the graph? What rate does the slope represent?

37 Write the equation for the graphed line in slope-intercept form.

38 Suppose the store increased the commission to 5% but decreased the weekly salary to $100. How would this change affect the graph? Write the equation for these new conditions.

The financial department of a production firm presented the graph shown below of the production costs for a certain item. The graph shows the fixed costs—the costs for tooling and overhead that are required to produce even one item—and the variable costs—the costs due to materials, labor, marketing, transportation, and so on.

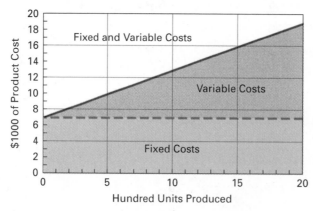

39 Identify the slope and *y*-intercept of the graph for total production cost.

40 How are these (slope and *y*-intercept) related to the fixed costs and the variable costs?

You must have some brochures printed, so you make a telephone call to a local print shop. You are told that there is no setup charge. The cost for printing various quantities is given to you in a table.

Quantity	Cost
50	$ 7.68
100	8.91
150	10.14
200	11.36
250	12.59
500	18.73
750	24.86
1000	31.00

41 Choose a scale and graph the cost data. Draw a line that seems to fit the data.

42 Determine the slope and *y*-intercept of your line.

43 What is the meaning of this nonzero *y*-intercept?

HEALTH OCCUPATIONS

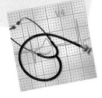

Research indicates that blood flow through the kidneys decreases with age, as shown in this simplified graph of experimental findings.

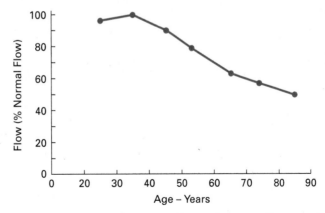

44 Would you say that the changing blood flow shows a linear relationship, or close to it? If not, could you say that it was linear for some ages?

45 A computer analysis of these data reported the following equation for relative blood flow.

$$F = -1.18A + 141$$

A is the age in years, from 35 to 85 years.

Evaluate the equation for several ages to see if it agrees with the graphed data. (Round your answers to the nearest whole number before comparing to the graph.)

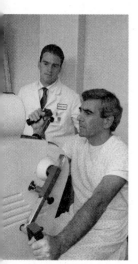

A "stress test" is a method of evaluating the health of a patient's cardiovascular system. In this test, a technician monitors the patient's pulse rate during an exercise session on a treadmill or stationary bicycle. A patient's pulse rate is not allowed to exceed a maximum rate. The maximum rate is based on the patient's age, from the following graph.

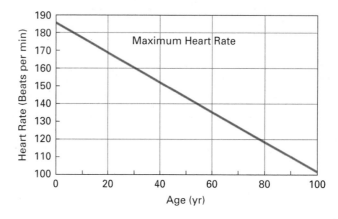

46 What is the *y*-intercept of the graph?

47 What is the slope of the graph?

48 Write an equation for the graphed line in the slope-intercept form.

49 Use your equation to find the maximum heart rate in a stress test for a 44-year-old patient.

Weight loss occurs when more calories are expended in exercise than are consumed in the daily diet. When 3500 calories more are used in exercise than are consumed in the diet, about 1 pound of fat is lost. The following graph shows the weight loss versus calories expended in exercise that can be expected by a 5 foot 6 inch, 132-pound woman who consumes a 2400-calorie daily diet. (Her weight remains stable with the 2400-calorie daily diet and normal exercise.)

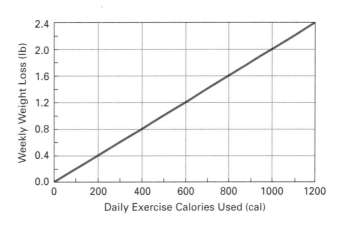

50 What is the slope of the line?

51 What is the equation of the line?

52 How much weight will the woman lose each week if her exercise uses 350 calories per day?

53 If the woman decreases her dietary calories from 2400 calories to 2150 calories per day, how will this affect the graphed line? What is the equation for the new line?

A pickle recipe requires $\frac{3}{4}$ ounce of salt per pound of cucumbers.

1 lb

$\frac{3}{4}$ oz

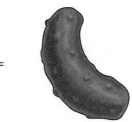

+

S

=

54 Write an equation in slope-intercept form that shows the relationship between the pounds of cucumbers and the ounces of salt.

55 How much salt do you need to pickle 1 ton of cucumbers? Remember that 1 ton = 2000 pounds.

56 How many pounds of cucumbers can you pickle with 100 pounds of salt? Remember that 1 pound = 16 ounces.

INDUSTRIAL TECHNOLOGY

When you use a thermocouple to measure the temperature at a location, you must convert a voltage reading from the thermocouple into a temperature. You recorded the following voltages from a voltmeter and the corresponding temperatures from a thermometer.

Temperature (°C)	Voltage (mV)
0	0.00
100	6.32
200	13.42
300	21.03
400	28.94
500	37.00
600	45.09

57 Graph the data. Based on your graph, is the relationship between temperature and thermocouple voltage a linear one? Explain your answer.

58 Write a formula for a line, in slope-intercept form, that will very closely describe the relationship between the two variables T (temperature in °C) and V (voltage in millivolts). This is the calibration equation for the thermocouple.

59 How can you use this equation for temperature measurements?

60 What is the meaning of the y-intercept in the calibration equation? What rate does the slope represent?

You have a standard assembly used to construct large exhaust fans for industrial warehouse facilities. The assembly is designed so that the motor pulley and the fan pulley are 28 inches apart. A small adjustment is available for belt tension. The motor pulley has a diameter of 3 inches. Based on the required fan speed and fan size, you can vary the size of the fan pulley. To help you determine the belt sizes, you have the formula

$$B = 2L + 1.625(D + d)$$

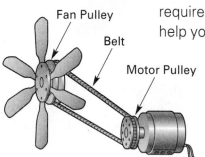

Fan Pulley
Belt
Motor Pulley

B is the length of the belt needed.
L is the distance between the two pulley centers.
D is the diameter of one pulley.
d is the diameter of the other pulley.

61 Substitute the known values given above into the formula and rewrite the formula as a linear equation in slope-intercept form. Identify the variables, the coefficient, and the constant. In linear equations, what are other names for the coefficient and the constant?

62 Try your equation for fan pulley diameters of 6 inches and 10 inches. Check your results with the original formula.

You can use a flow test to determine if a radiator is clogged. In a flow test, you measure the volume of water that flows through a radiator by gravity in a measured time period. Automotive manufacturers establish the standards for these tests. If less water than the standard flows through a radiator, it is clogged. The standard established by Ford Motor Company is 42.5 gallons per minute.

63 Write an equation in slope-intercept form that shows the relationship between the time in seconds and the volume in gallons measured by the flow test.

64 Use the equation to calculate the water that should flow through a Ford radiator that is not clogged in 20 seconds.

65 Use the equation to calculate the time needed for a Ford radiator that is not clogged to fill a 5 gallon bucket.

66 Suppose you test a Ford radiator and measure a flow of 25 gallons of water in 40 seconds. Should you consider the radiator clogged?

The following table gives selected values of wire gauge numbers and the corresponding wire diameters.

Wire Gauge	Diameter (in inches)
0	0.3249
5	0.1819
10	0.1019
15	0.05707
20	0.03196
25	0.01790
30	0.01003
35	0.005615
40	0.003145

67 Draw a graph of the data.

68 Is the relationship between wire gauge number and diameter a linear relationship?

CHAPTER 5   ASSESSMENT

Skills

Write the number pair represented by each point on the graph below.

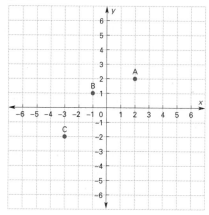

1. Point A　　　　　**2.** Point B　　　　**3.** Point C

Name the quadrant where each point is graphed.
4. $(-4,-3)$　　　**5.** $(2,-6)$　　　**6.** $(-1,7)$

Find the slope of the line through each pair of points.
7. $(-2,3)$ and $(5,2)$　**8.** $(2,4)$ and $(7,-1)$　**9.** $(6,0)$ and $(0,6)$

Name the slope and y-intercept of the graph for each equation.
10. $y = 3x + 7$　　**11.** $y = -2x + 5$　　**12.** $y = \frac{2}{3}x - 4$

Write each equation in slope-intercept form.
13. $x + 2y = 6$　　**14.** $2x - y + 5 = 0$　**15.** $3y - 5x = 6$

Applications

16. If $(2,3)$, $(-2,3)$, and $(2,-3)$ are vertices of a rectangle, what are the coordinates of the fourth vertex? Explain your answer.

17. A department store sells suit jackets and skirts. The jackets are always twice as much as the matching skirt. Write an equation to represent the relationship of jacket cost to skirt cost.

18. Mr. Garcia drives at a constant speed for 5 hours. At the end of two hours, he had driven 90 miles. After four hours, he had driven 180 miles. How many miles had he driven after five hours? Write an equation and draw a graph to determine your answer. At what rate is Mr. Garcia driving?

19. A cable TV company charges $30 for a basic installation plus $5 for each additional television it hooks up. If the total cost is a linear function, write an equation to represent this function and find the cost of having a cable installation including four additional televisions.

20. A company has developed a new product. Market research has determined that the demand for the product is a linear function of the price. The company can sell 70 items per month if it charges $150 each, but only 20 items if it charges $200 each. How many items of the new product will the company sell if it charges $180 each?

Math Lab

21. The volume of a cylinder is $V = \pi r^2 h$. If you graph the volume of water in a can as a function of the height of the water, you see a straight line. What is the slope of the line?

22. If one inch is equal to $2\frac{54}{100}$ centimeters, write a linear equation to convert centimeter measures to inches. What is the y-intercept of your equation?

23. You use a graduated cylinder to measure the volume and weight of various amounts of water. When you plot the weight on the horizontal axis and the volume on the vertical axis, you see a straight line. But the line does not pass through the origin—there is a y-intercept. What does the y-intercept represent?

CHAPTER 6

WHY SHOULD I LEARN THIS?

The next wave of technological breakthroughs will involve nonlinear systems. Industry needs people who can use nonlinear mathematical models to create new products that will improve our lives. This chapter shows how nonlinear systems are applied to the workplace and are important to your future.

NONLINEAR FUNCTIONS

OBJECTIVES

1. Represent relations and functions as tables of data, ordered pairs, graphs, and equations.
2. Identify the domain and range of a function.
3. Identify and graph nonlinear functions involving absolute value, squares, square roots, exponents, and reciprocals.
4. Solve problems involving nonlinear equations either by graphing or by using algebraic methods.

In chapter 5 you studied the patterns of linear functions. You represented these patterns with tables of ordered pairs, with graphs, and with equations of the form $y = mx + b$.

In this chapter you will see how patterns of nonlinear functions can also be represented by tables, graphs, and equations.

Automotive engineers and technicians use nonlinear functions to design and test electric cars. Structural engineers model the motion on bridges with nonlinear functions.

As you watch the video, look for graphs that are not straight lines. These graphs will represent nonlinear functions.

LESSON 6.1 RELATIONS AND FUNCTIONS

You use relationships between two variables every day. One example of a *relation* is the phrase "is a friend of."

In the sentence,

x is a friend of *y*

any two friends can replace the variables.

Jamie is a friend of Roberto

is an example of the relationship.

Sets of ordered pairs, tables, and graphs are all useful tools used to represent relationships between two variables. For example, you can estimate the distance of an approaching thunderstorm by counting the seconds between a flash of lightning and the resulting sound of thunder. This works because sound travels about $\frac{1}{3}$ of a kilometer every second. Consequently, if 3 seconds go by between the flash and the sound, the storm is 1 kilometer away, and so on.

A set of ordered pairs is called a **relation**. You can use ordered pairs to represent the data for an approaching thunderstorm.

$$R = \{(3, 1), (6, 2), (9, 3), (12, 4), \ldots\}$$

The braces { } mean *the set of.* You can list a relation in a table or draw its graph on a coordinate plane.

Time in Seconds	Distance in Kilometers
3	1
6	2
9	3
12	4
⋮	⋮

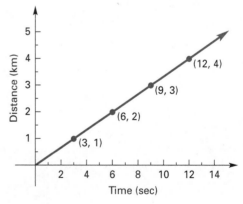

There is a pattern in the ordered pairs describing the thunderstorm data. Represent time with the independent variable t and the distance with the dependent variable d. The following equation represents the relationship:

$$d = \frac{1}{3}t$$

This equation is a shortcut way of saying that *the distance sound travels depends on time*.

Sequences

Examine the following sequence of numbers:

$$88, 96, 104, 112, 120, \ldots$$

The numbers in this sequence record the speed of an electric car each time it passes the timing line. The first speed measured is 88 feet per second and the car is steadily increasing its speed. Do you see a pattern? The difference between any two numbers is 8. You can order the numbers in the sequence with the set of whole numbers and list them in a table.

Lap Number	Speed	Difference in Speed
0	88	
1	96	$96 - 88 = 8$
2	104	$104 - 96 = 8$
3	112	$112 - 104 = 8$
4	120	$120 - 112 = 8$

If you graph the ordered pairs, you can see that the graph is a line with slope 8 and y-intercept 88. The equation representing the relationship between the ordered pairs is

$$y = 8x + 88$$

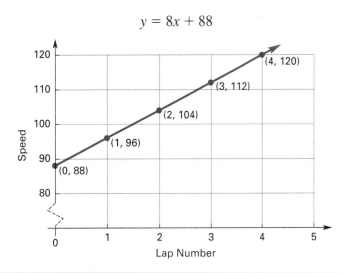

Ongoing Assessment

What is the speed of the car after 5 laps? By how much does the car increase its speed each lap?

EXAMPLE 1 Charges for Service Calls

The XYZ Plumbing Company has a schedule of charges for making service calls. The company charges a flat rate of $40 for each call plus $20 per hour on the job. Show the schedule of charges as a table, a graph, and an equation.

SOLUTION

Time in Hours	Charge in Dollars
0	40
1	60
2	80
3	100
4	120

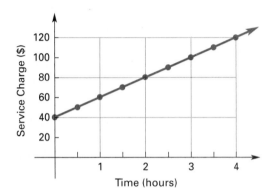

The equation $c = 20t + 40$ represents the relationship between the time worked (t) and the charge for the work (c).

ACTIVITY   Sequences and Equations

Use the sequence 3, 7, 11, 15, 19, . . . to complete the following:

1 Use whole numbers to write the sequence as a set of ordered pairs beginning with (1, 3) and ending with (5, 19).

2 Use the ordered pairs to make a table.

3 Graph the set of ordered pairs.

4 Find the slope and y-intercept of the graph.

5 Find an equation that expresses the relationship between the numbers in the sequence.

6 Extend the sequence to three more numbers.

Function Notation

A **function** is a relation with one additional condition.

> **Function**
>
> A function is a set of ordered pairs (x, y) such that for any value of x, there is exactly one value of y.

A function machine can explain how a function operates. If you put in a certain sequence of numbers, you will get out a definite sequence of numbers. Sometimes you can determine the rule by examining the input and output numbers.

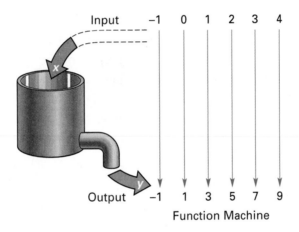

Function Machine

The equation $y = 2x + 1$ describes the rule used in the function machine. In this case, x is the independent variable and y is the dependent variable. Since a *unique* value for y is determined for each value of x, y is a function of x. A relation is a function if for each value of the independent variable x, there exists exactly one value of the dependent variable y. In function notation, you write

$$f(x) = 2x + 1$$

The symbol $f(x)$ is read "f of x" or "f is a function of x." In function notation, an input x determines the output $f(x)$ according to the rule $2x + 1$. Usually the letters $f, g,$ and h are used to represent functions. The set of input values is the **domain** of the function. The set of output values is the **range**.

EXAMPLE 2 Evaluating a Function

Let $f(x) = 3x - 2$. What is the value of $f(x)$ when $x = 5$?

SOLUTION

Substitute 5 for x in the rule and simplify.

$$f(5) = 3(5) - 2$$
$$= 15 - 2$$
$$= 13$$

Thus, $f(5) = 13$.

Ongoing Assessment

a. If $h(x) = -5x - 6$, what is $h(-4)$?

b. If $g(x) = x^2 - 5x + 6$, find $g(2)$ and $g(3)$.

When you use function notation where $y = f(x)$, graph x on the horizontal axis and $f(x)$ on the vertical axis. The graph of the function in Example 2 looks like this.

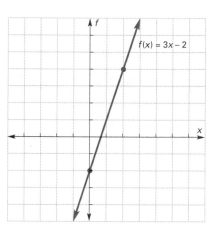

LESSON ASSESSMENT

Think and Discuss

1 Give examples of mathematical relations that are not functions. Explain why they are relations but not functions.

2 How many ways can you describe a function? Give examples.

3 Explain why a sequence of numbers might be considered a function.

4 Explain what the notation $g(x) = 3x + 5$ means.

5 Explain how to find the value of a function.

For each sequence, write a rule that gives the relationship between the numbers and find the next three numbers in the sequence.

6. 3, 4, 5, 6

7. 0, 2, 4, 6

8. −1, 2, 5, 8

9. 6, 11, 16, 21

10. −1, 1, 3, 5

11. 1, $\frac{3}{2}$, 2, $\frac{5}{2}$

If $h(x) = 2x - 7$, find

12. $h(4)$

13. $h(0)$

14. $h(-3)$

If $f(x) = -5x + 3$, find

15. $f(0.2)$

16. $f(-1)$

17. $f\left(\frac{3}{5}\right)$

For each situation, make a table, find and graph the equation, and give the slope and y-intercept.

18. Roberto started walking from a point 3 miles from his house and walked away from his house at a constant speed of 4 miles per hour. What is Roberto's distance (d) from home as a function of time (t)?

19. In Carlette's appliance repair business, she charges $15 for making a house call. She also charges $8 per hour. What are Carlette's total charges (c) as a function of time (t)?

20. Videos rent for $3 per day. To join the video club, you pay a one-time fee of $12. What is the cost ($y$) of renting videos for x days?

21. Write 0.000085 in scientific notation.

22. Write 100,000,000 as a power of ten.

Solve each equation.

23. $5y - 4 = 25$

24. $2x + 9 = -7$

25. $3(r - 1) = -9$

26. 20% of $g = 20$

27. $95 = 15x + 20$

28. $\dfrac{m}{-5} = \dfrac{16}{-10}$

29. A store pays $650 for a stereo system. It charges its customers $1050 for the system. What is the profit from the sale of a stereo system written as a percent of the store's cost?

LESSON 6.2 THE ABSOLUTE VALUE FUNCTION

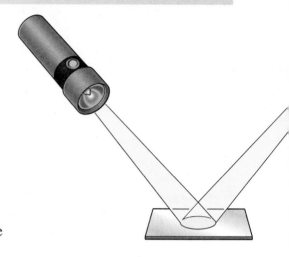

In the "function machine" illustrated in Lesson 6.1, each input number resulted in exactly one output number. In some functions, however, two or more input numbers might result in the same output number. The absolute value function is one such function.

Recall that the absolute value of a number is its distance from zero on the number line. For example, $|-2| = 2$ and $|2| = 2$. Both -2 and 2 are two units from zero on the number line. This leads to a definition of absolute value expressed with variables.

Absolute Value

If $x \geq 0$, then $|x| = x$

If $x < 0$, then $|x| = -x$

The absolute value function is written

$$y = |x| \qquad \text{or} \qquad f(x) = |x|$$

The domain of this function is the set of all numbers, but the range is restricted to the set of positive numbers and zero.

Critical Thinking   Why are functions that involve the measurement of time or distance usually restricted? Describe a function that has a restricted domain and range. In which quadrants does its graph belong?

EXAMPLE 1 Graphing the Absolute Value Function

Graph $y = |x|$.

SOLUTION

1. Make a table.

x	$|x|$
-2	2
-1	1
0	0
1	1
2	2

2. Draw a graph.

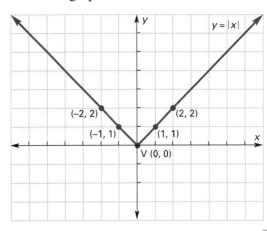

The graph of the absolute value function looks like a V. The point V(0,0) is called the **vertex** of the graph.

Critical Thinking   Compare the graph of the absolute value function $y = |x|$ with the graphs of $y = x$ and $y = -x$. How are the graphs alike and different?

A graphics calculator will make the following activities easier to do. Check your calculator to see how to graph the absolute value function. If you do not have access to a graphics calculator, make a table for each function and draw its graph.

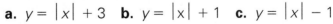

ACTIVITY 1   Vertical Slides

1 Graph $y = |x|$.

2 Graph each function on the same coordinate axes.

   a. $y = |x| + 3$ **b.** $y = |x| + 1$ **c.** $y = |x| - 1$

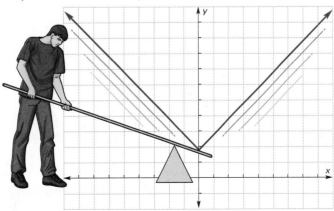

3 The function $y = |x|$ is called the **parent function** for absolute value. Each of the other absolute value graphs results from a movement of the parent graph. Explain how adding or subtracting a constant value can affect the graph of the parent function.

When the graph of a parent function slides up or down, the slide is called a **vertical translation**.

> **Vertical Translation**
> If $y = |x| + c$,
>
> then the vertex of $y = |x|$ is translated up or down by c units.

ACTIVITY 2 Horizontal Slides

1 Graph $y = |x|$.

2 Graph each function on the same coordinate axes.

  a. $y = |x - 3|$ **b.** $y = |x + 1|$ **c.** $y = |x - 1|$

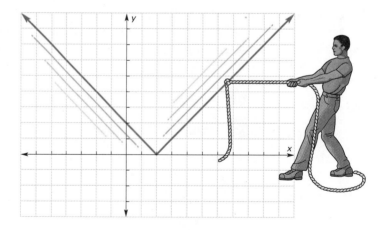

3 What is the effect of
a. adding a constant within the absolute value bar?

b. subtracting a constant within the absolute value bar?

When the graph of a parent function slides right or left, the slide is called a **horizontal translation**.

> **Horizontal Translation**
> If $y = |x + c|$,
>
> then the vertex of $y = |x|$ is translated right or left by c units.

Ongoing Assessment

Describe the graph of the function $h(x) = |x + 2| - 5$.

ACTIVITY 3 Stretching and Shrinking

1 Graph each function on the same coordinate axes.

    a. $y = 2|x|$ **b.** $y = 3|x|$ **c.** $y = 5|x|$

2 Compare each function to $y = |x|$.

3 Graph each function on the same coordinate axis.

    a. $y = \frac{1}{2}|x|$ **b.** $y = \frac{1}{3}|x|$ **c.** $y = \frac{1}{5}|x|$

4 Compare each function to $y = |x|$.

5 What effect does the constant a in $y = a|x|$ have on the parent function when a is less than 1 and greater than zero?

When the parent function $y = |x|$ is multiplied by a constant, the constant is called the **scale factor** of the function.

What happens when each value of $y = |x|$ is multiplied by the scale factor 3? Each point of each arm of the V is moved up by a factor of 3. This has the effect of pulling the arms of the V together. In 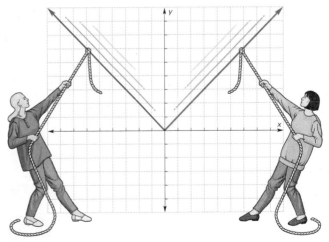 other words, the angle between the arms has shrunk. On the other hand, if the scale factor is $\frac{1}{2}$, each point on each arm of the V is moved down by a factor of $\frac{1}{2}$.

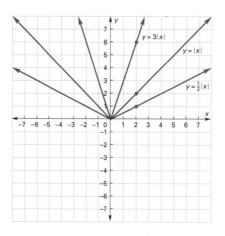

Scale Factor

If $y = a|x|$, then a is the scale factor of the function.

If a is greater than 1, the angle between the lines of the graph becomes smaller or shrinks.

If a is greater than zero, but less than 1, the angle becomes greater or stretches.

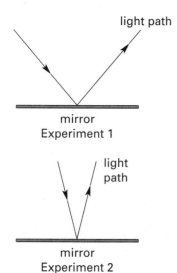

light path

mirror
Experiment 1

light path

mirror
Experiment 2

A technician is performing an experiment with a laser. He is aiming the beam at a mirror and checking its reflection.

Can the technician use an absolute value function to describe the path of the light?

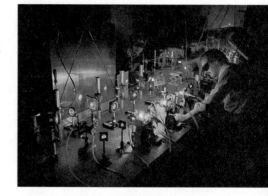

Suppose the laser is moved to another position to perform another experiment. How can the absolute value function described in the first experiment represent the light path in the second experiment?

EXAMPLE 2 Combining Moves

Describe the graph of $f(x) = 4|x| - 1$.

SOLUTION

The scale factor 4 shrinks the angle between the lines of the graph of the parent function by 4 units. Since the constant 1 is subtracted, the graph is moved down 1 unit.

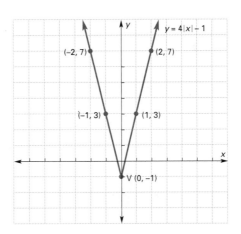

$y = 4|x| - 1$

$(-2, 7)$ $(2, 7)$

$(-1, 3)$ $(1, 3)$

V $(0, -1)$

ACTIVITY 4 A Reflection

1 Repeat Activity 3 with each scale factor changed to a negative number.

2 Explain what happens to the parent graph under these circumstances.

3 Create a definition for reflecting the absolute value function over the *x*-axis.

4 Check with your classmates. How do your definitions compare?

WORKPLACE COMMUNICATION

Use the absolute value function to model the struts at the current position and at the suggested new position. What scale factor can be used for the absolute value function for the new position of the struts? If the suggestion is implemented, what length of material is needed to fabricate the new struts?

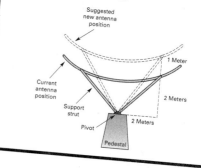

Modern Workplace, Inc.
SUGGESTION PROGRAM

INSTRUCTIONS: Submit this form to your supervisor or the Human Resources Office.
REMINDER: If your suggestion for improved processes or products is accepted and implemented by Modern Workplace, Inc. you will receive 50% of the monetary savings for the first year of implementation. If your suggestion for improved safety or security is implemented, you will receive a one-time $500 cash award.

Your Name(s): Roberta Duggan
Office: Communications Support

email Address: duggar@mwi.org
Extension: 6-4044

Describe your suggestion for improvement:

We are frequently tasked with installing microwave electronic components in the MWI satellite antenna for repair or development testing. To work in the middle portion of the antenna, we can tilt it right or left on a pedestal mounted pivot. However, the support struts limit how far the antenna can be tilted. Currently, it will not tilt far enough to allow access to the middle. When we cannot reach the middle, we have to completely disassemble the antenna to work on the electronics.

I recommend the antenna be remounted on longer support struts (see diagram) so that the antenna is raised one meter from its current position. In this new position, it would tilt enough to the right or left to allow access to the electronics at the middle. This will save at least six hours (for a two person crew) each time we avoid disassembling and reassembling the antenna.

Suggested
new antenna
position

1 Meter

Current
antenna
position

2 Meters

Support
strut

2 Meters

Pivot

Pedestal

LESSON ASSESSMENT

Think and Discuss

1 Explain why $y = |x|$ is a function.

2 What happens to the vertex of $y = |x|$ when 3 is added to or subtracted from the function? Give an example.

3 What happens to the shape of $y = |x|$ when the function is multiplied by 10? Give an example.

4 What happens to the shape of $y = |x|$ when it is multiplied by $\frac{1}{4}$? Give an example.

5 Explain what happens to the graph of $y = |x|$ when it is multiplied by a negative scale factor.

Practice and Problem Solving

Graph each function. Describe the graph in relation to the parent function $y = |x|$.

6. $y = |x| + 7$

7. $h(b) = 2|b|$

8. $r = \frac{3}{4}|t|$

9. $g(s) = -3|s| + 1$

10. $y = \frac{1}{2}|x| - 3$

11. $a = 4|b| + 6$

12. $f(x) = \frac{2}{3}|x + 4|$

13. $b = 3|c| + 2\frac{1}{2}$

14. $y = -\frac{5}{2}|x| - 2$

Mixed Review

15. The science club treasury had a balance of $350.45. Over the next four weeks, the following withdrawals (W) and deposits (D) are made. Find the final balance.

$114.36(W); $39.57 (W); $205.15(D); $185.59 (W)

Write each expression as a power of ten.

16. 10,000,000

17. 0.00001

Solve each equation.

18. $5a - 9 = 26$

19. 30% of $150 = c$

20. $\frac{-6}{9} = \frac{m}{15}$

21. 6% of $q = 96$

22. $\frac{3}{4}x - 7 = -19$

23. $7 = 0.5x - 3$

Write and solve an equation for each situation.

24. The yearly simple interest rate paid on an investment is 7.5%. If $15,000 is invested, what is the interest paid for two years?

25. One angle of a triangular support is three times as large as the smallest angle. The largest angle is 80° greater than the smallest angle. Find the measurement of each angle.

Find the slope and *y*-intercept.

26. $3y - 9 = 5x$ **27.** $3x + 2y = 12$

LESSON 6.3 THE QUADRATIC FUNCTION

Area of a Square

Is the equation $y = x^2$ a function? To find out, first make a table. Then draw a smooth curve connecting the points.

x	x^2
-2	4
-1	1
0	0
1	1
2	4

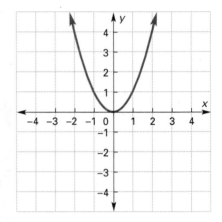

Note that for each input value in the domain, there is exactly one output value in the range. The equation $y = x^2$ is a function. This function models the area of a square. Replace the dependent variable (y) with the symbol A to represent area. Replace the independent variable (x) with the symbol s to represent the length of a side. The equation for the area of a square is

$$A = s^2$$

Critical Thinking What, if any, are the restrictions on the domain and range of the function $A = s^2$?

The Parabola

An equation written in the form $y = x^2$ or $f(x) = x^2$ is called a **quadratic equation** or **quadratic function**. A quadratic equation or function has exactly one term in the expression that is raised to the second power. The area formula is an example of a quadratic function. The graph of a quadratic function is a curve called a **parabola**. The point V(0,0) is the **vertex** of the parabola. Many real world objects are described by parabolas.

Quadratic functions can model the relationship between the distance (d) and time (t) for a free-falling object.

EXAMPLE 1 Quadratic Functions

How far will a bungee jumper free-fall in 5 seconds? What are the domain and range for the function?

SOLUTION

The function is $d = 16t^2$, where d is in feet and t is in seconds. Evaluate the function when t is 5 seconds. Thus, the distance is $d = 16(5^2)$ or 400 feet. The domain is measured in time and the range in distance. Thus, the domain and the range are both restricted to numbers greater than or equal to zero.

Ongoing Assessment

Graph the function $y = x^2$. Describe the graph.

Critical Thinking   Compare the graph of $y = x^2$ with $y = |x|$. How are they alike and different?

ACTIVITY 1   Vertical Translations

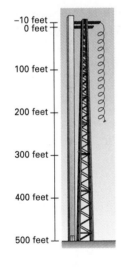

-10 feet
0 feet
100 feet
200 feet
300 feet
400 feet
500 feet

1 Graph each function on the same coordinate axes.

   a. $y = x^2 + 3$ **b.** $y = x^2 + 1$ **c.** $y = x^2 - 1$

Describe each graph.

2 Describe how the constant c in $y = x^2 + c$ alters the graph of the parent function $y = x^2$.

3 Compare $y = |x| + c$ with $y = x^2 + c$. Use numbers to check your comparison.

4 Suppose in the bungee jumper problem another person jumps from a platform that is 10 feet higher. Explain why the equation $16t^2 - 10$ describes the motion of the second bungee jumper.

ACTIVITY 2   Horizontal Translations

1 Graph each function on the same coordinate axes.

   a. $y = (x - 2)^2$ **b.** $y = (x + 3)^2$ **c.** $y = (x - 1)^2$

Describe each graph.

2 How does the constant c in $y = (x + c)^2$ alter the graph of the parent function $y = x^2$?

3 How are $y = |x + c|$ and $y = (x + c)^2$ alike?

ACTIVITY 3   Stretching and Shrinking a Parabola

1 Graph each function on the same coordinate axes.

       a. $y = 3x^2$ **b.** $y = 2x^2$ **c.** $y = \frac{1}{2}x^2$

Describe each graph.

2 Describe how the constant a in $y = ax^2$ alters the graph of the parent function $y = x^2$ when a is greater than 1. What happens when a is greater than zero, but less that 1?

3 The moon's gravitational force is only about $\frac{1}{6}$ that of the Earth. If a ball is dropped on the moon, the equation that models the ball's motion is $d = 2.67t^2$. Explain how the constants in the equations $16t^2$ and $2.67t^2$ affect the motion of the ball.

ACTIVITY 4   Reflecting a Parabola

1 Graph each function on the same coordinate axes.

       a. $y = -3x^2$ **b.** $y = -2x^2$ **c.** $y = -\frac{1}{2}x^2$

Describe each graph.

2 Describe how the constant a in $y = ax^2$ alters the graph of the parent function $y = x^2$ when a is less than zero.

The graphs of $y = x^2$ and $y = -x^2$ are mirror images of one another. The graph of $y = x^2$ has been flipped over the x-axis to graph $y = -x^2$. This flip is called a **vertical reflection**.

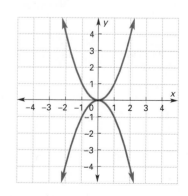

Vertical Reflection

The graph of $y = -x^2$ is called the vertical reflection of $y = x^2$ and appears as the mirror image of $y = x^2$ across the x-axis.

EXAMPLE 2 Moving a Parabola

Graph the function $y = -2x^2 + 1$.

SOLUTION A Without a Graphics Calculator

Graph $y = x^2$.

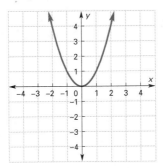

Shrink by the scale factor 2.

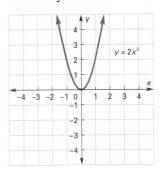

Reflect it over the x-axis.

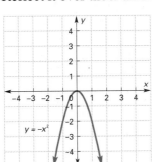

Move the vertex up 1 unit.

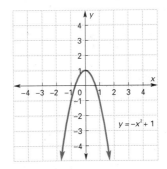

SOLUTION B With a Graphics Calculator

Set an appropriate Range.

Input the function.

Press the (GRAPH) key.

Critical Thinking Explain how to move the graph of the parent function $y = |x|$ to graph $y = -\frac{1}{2}|x| - 3$.

EXAMPLE 3 Braking Distance

A driver education course teaches that the higher the speed of a car, the greater the distance needed to stop. You might think that if you go twice as fast, you need twice the distance to stop. Is this true? Let b represent the braking distance in feet. Let v represent the speed of the car in miles per hour when you start to brake. The function

$$b = 0.037\, v^2$$

is the estimated braking distance of a car on dry concrete. Make a table of car speed and braking distance. Examine the pattern in the table. If a car moves twice as fast, does it need twice the distance to stop?

SOLUTION

The function is a quadratic. The domain and range are restricted to numbers greater than zero. The scale factor (0.037) will horizontally stretch the parent function. Thus, the faster you drive, the more distance you need to brake. You can illustrate this by graphing the ordered pairs formed by speeds of 20, 30, 40, 50, and 60 miles per hour.

Speed (mph)	Braking Distance (ft)
20	15
30	33
40	59
50	93
60	133

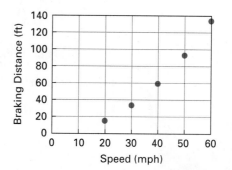

The braking distances are rounded to the nearest whole number to simplify the graph. You can see that braking distance does not double as you double your speed.

WORKPLACE COMMUNICATION

What is the "unstretched" length of the bungee cord you should use for the tests? (Recall the quadratic function $d = 16\,t^2$ for the number of feet an object freely falls in t seconds.)

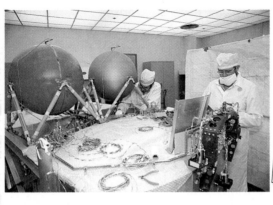

LESSON ASSESSMENT

Think and Discuss

1 Explain why $y = x^2$ is a function.

2 What happens to the vertex of the parent quadratic function when you add or subtract a positive constant?

3 What happens to the vertex of the parent quadratic function

a. when you add a positive constant (c) to the independent variable?

b. when you subtract a positive constant (c) from the independent variable?

4 What happens when you multiply the parent quadratic by a constant greater than 1?

5 What happens when you multiply the parent quadratic by a constant greater than zero but less than 1?

6 Explain what happens when you multiply the parent by a number less than zero.

Practice and Problem Solving

The power, in watts, of a microwave transmitter is given by the function $P = I^2R$. I is the electrical current passing through the transmitter in amperes, and R is the resistance of the transmitter in ohms. Find the power of a 4-ohm transmitter when the current is

7. 5 amps   **8.** 10 amps   **9.** 20 amps   **10.** 30 amps

Let $g(x) = -3(x + 2) - 4$. Find $g(x)$ when x is
11. -2 **12.** 4   **13.** 0   **14.** $-\dfrac{1}{3}$

Graph each function. Describe the graph in comparison to the parent graph $y = x^2$.
15. $y = x^2 - 1$ **16.** $g(a) = -a^2 + 3$ **17.** $d = 3(t + 2)^2 + 1$

18. $h(n) = -2n^2 - 3$ **19.** $A = \pi r^2$ **20.** $f(x) = \dfrac{1}{2}x^2 - 5$

Mixed Review

At 4 AM the temperature is $-8°F$. At 11 AM it is $-12°F$.

21. At what time is the temperature greater?

22. What is the difference in temperature between the two readings?

23. During the next three hours, the temperature increases 5 degrees. What is the new temperature?

24. Convert the new temperature to Celsius.

Write an equation and solve.
25. What percent of 80 is 15?   **26.** 35 is what percent of 10?

Solve each equation.
27. $4a - 6 = 9$ **28.** $3(2x - 3) = -5$

Graph each equation.
29. $2y + x = 4$ **30.** $3y - 5x = 8$

LESSON 6.4 THE SQUARE ROOT FUNCTION

You have already used the formula $A = s^2$ to find the area (A) of a square when the length of a side (s) is known. If you know the area, the following formula provides a means for finding the length of one side:

$$s = \sqrt{A}$$

This formula is read "s is equal to the square root of A." The equation $s = \sqrt{A}$ is an example of the general equation $y = \sqrt{x}$.

What is the meaning of $y = \sqrt{x}$? Since $2^2 = 4$, 2 is a square root of 4. Thus, $\sqrt{4} = 2$. However, $(-2)^2 = 4$. Thus, $\sqrt{4} = -2$. Every positive number except zero has two square roots. Since $\sqrt{4} = 2$ or -2 (written as $\sqrt{4} = \pm 2$), one input number (4) has two output numbers (2 or -2). Thus, the equation $y = \sqrt{x}$ does *not* represent a function. To find out more about this equation, you can make a table and draw a graph.

x	$\sqrt{x}$
0	0
1	± 1
4	± 2
9	± 3

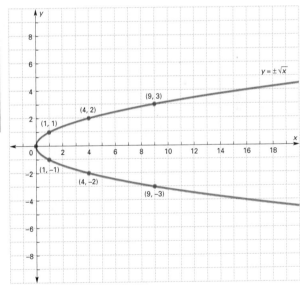

You can tell by the table that "finding the square root of a number" is not a function. However, if you look at the graph, you can see that the top curve represents positive square roots and the bottom curve represents negative square roots. Furthermore, the curves are vertical reflections of each other across the x-axis. If you restrict the range to positive values and zero, $y = \sqrt{x}$ becomes a function. This function is the principal square root function. From now on *the square root sign will mean positive square root.* If the negative square root is needed, you write $-\sqrt{}$.

> **Square Root**
> If $x \geq 0$, $\sqrt{x^2} = x$.

Solving Equations

We have now defined $y = \sqrt{x}$ as a function.

Critical Thinking Explain how to find the domain and range of the function $f(x) = \sqrt{x}$.

The equation $y = \sqrt{4}$ has exactly one solution. The solution is 2. However, the equation $x^2 = 4$ has two solutions, 2 and -2.

> **Solving $x^2 = c$**
> If $x^2 = c$ and $c \geq 0$,
> 1. $x = \pm\sqrt{c}$.
> 2. The solutions are $\sqrt{c}$ and $-\sqrt{c}$.

EXAMPLE 1 Falling Objects

A ball falls from the top of a 250-foot building. The equation that models the distance (d) it falls in time (t) is $d = 16t^2$. How long does it take the ball to hit the ground?

SOLUTION

Let d equal 250 and solve the equation for t.

$16t^2 = 250$	Given
$t^2 \approx 15.6$	Division Property (calculator)
$t \approx \pm\sqrt{15.6}$	Solving $x^2 = c$

Since t represents time, only the positive square root makes sense. Using your calculator, you will find that it takes about 3.9 seconds for the object to reach the ground.

The Allied Storage Company has a new storeroom that provides 6000 square feet of floor space. If the storeroom has a square shape, find the approximate length of each side.

The Pythagorean Theorem

A **right triangle** contains exactly one right angle. Many early civilizations found an interesting relationship between the three sides of a right triangle.

leg — hypotenuse — leg

If you place squares on each of the sides of a right triangle and measure the area of each square, you will find that

the area of the square on the hypotenuse = the sum of the areas of the squares on the other two sides.

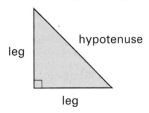

4 in.   16 in.²   *b*   *c*   25 in.²   5 in.

a

9 in.²

←3 in.→

This relationship is called the **Pythagorean Theorem** or Pythagorean Formula. (To find out why, read the Cultural Connection at the end of this Lesson.) The Pythagorean Theorem is usually written in the form

$$c^2 = a^2 + b^2$$

In this form, c is the length of the hypotenuse, and a and b are the lengths of the legs. Applying the Pythagorean Theorem to the right triangle in the illustration gives

$$c^2 = a^2 + b^2$$

$$5^2 = 3^2 + 2^2$$

$$25 = 9 + 16$$

EXAMPLE 2 Finding Distance

An escalator takes you from the second floor to the third floor of a building. During the ride, the escalator rises a distance of 30 feet and moves horizontally a distance of 27 feet. What is the length of the escalator?

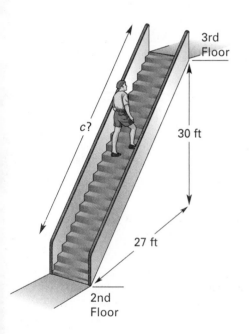

SOLUTION

Use a right triangle to model the situation. Let c represent the length of the escalator.

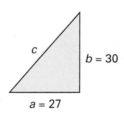

Use the Pythagorean Theorem. Substitute 27 for a, 30 for b, and solve for c.

$c^2 = a^2 + b^2$	Given
$c^2 = (27)^2 + (30)^2$	Substitution
$c^2 = 729 + 900$	Simplify
$c^2 = 1629$	Simplify
$c = \pm \sqrt{1629} \approx \pm 40.36$	Solve for c

Since only the positive square root makes sense, the escalator is about 40.36 feet in length.

You can write the Pythagorean Theorem in terms of a^2 and b^2.

Thus,

$$a^2 = c^2 - b^2 \qquad \text{and} \qquad b^2 = c^2 - a^2$$

EXAMPLE 3 Finding a Leg of a Right Triangle

Suppose the hypotenuse of a right triangle is 20 feet long. If one of the legs is 16 feet long, what is the length of the other leg?

SOLUTION

Use the formula $a^2 = c^2 - b^2$. Replace c with 20 and b with 16.

$$a^2 = (20)^2 - (16)^2$$

Use your calculator to find the length is 12 feet. Check your answer to be sure.

CULTURAL CONNECTION

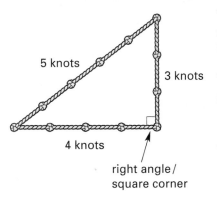

5 knots

3 knots

4 knots

right angle/
square corner

12 knots

The Pythagorean Theorem, in one form or another, was known by the ancient Chinese, Babylonians, and Egyptians. The Babylonians and Egyptians probably used the rule to survey lands after the flooding of their rivers. People also used the rule along with a knotted rope to mark off a square corner.

The earliest known informal proof of the rule is a diagram found in a Chinese manuscript called the *Chiu Chang*. However, a formal proof was not developed until many years later. During the sixth century BCE, a secret society called the Pythagoreans gathered to study philosophy, music, and mathematics. It is not known if the members of the society actually devised a formal proof, but the formula is named in honor of Pythagoras, the founder of this secret group.

LESSON ASSESSMENT

Think and Discuss

1 Explain why every positive number can have two square roots.

2 What is meant by the principal square root?

3 Why is $f(x) = \sqrt{x}$ considered a function?

4 Explain how to find the solution to $3x^2 = 48$.

5 Explain how to use the Pythagorean Theorem to find the length of the hypotenuse of a right triangle.

Practice and Problem Solving

Use your calculator to find each square root. If necessary, round to the nearest hundredth.

6. $\sqrt{225}$ **7.** $-\sqrt{2500}$ **8.** $-\sqrt{196}$ **9.** $\sqrt{615}$

Solve each equation. If the answer is not an integer, round it to the nearest hundredth.

10. $5a^2 = 500$ **11.** $2d^2 = 48$ **12.** $3x^2 - 8 = 100$

Use the Pythagorean Theorem to find the missing dimension of the triangle. If the answer is not an integer, round it to the nearest hundredth.

	side A	side B	hypotenuse
13.	3 meters	4 meters	?
14.	5 inches	7 inches	?
15.	?	12 feet	13 feet
16.	11 feet	35 feet	?
17.	21 inches	?	230 inches
18.	2 kilometers	5 kilometers	?

19. You ride your bike north 8 miles. Then you turn west and ride 7 miles. To ride directly home, how far do you have to travel?

20. A ladder is leaning against a wall at a point 30 feet above the ground. The bottom of the ladder is 40 feet from the wall. What is the length of the ladder?

21. You need to bury a water pipe in your backyard. Use the dimensions in the drawing to find what length pipe to buy.

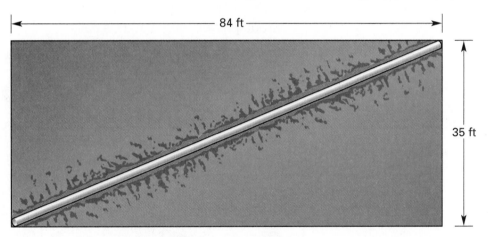

84 ft

35 ft

22. A baseball infield is shaped like a square. It is 90 feet between bases. How far is a throw from third base to first base?

Mixed Review

23. What is the value of $x(y + z)$ when x is -3, y is -5, and z is 8?

24. What is $f(-3)$ when $f(t)$ is $5t^2 - 3$?

25. What are the slope and y-intercept of $4x + y = -9$?

26. A manufacturer of lawn watering systems produces a sprinkler head that covers a circular area 12 feet in diameter. How many square feet of lawn should the manufacturer advertise the sprinkler head will cover?

27. A rectangular tank is 6 feet long, 8 feet wide, and 5 feet high. The tank is completely full of water. If one cubic foot is equivalent to about 7 gallons of water, about how many gallons of water will the tank hold?

Write and solve an equation for each situation.
28. Sharon and Jack are washing the windows in a large apartment building. Sharon has washed 36 more windows than Jack. If 148 windows have been washed so far, how many windows has Sharon washed?

29. The Top Notch Marketing Group spent three times as much on long distance calls as on local calls. The total phone bill is $415.36. How much did the Top Notch Marketing Group pay for long distance calls?

LESSON 6.5 EXPONENTIAL FUNCTIONS

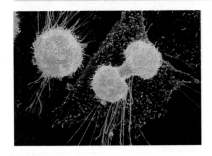

Plants and animals grow through a process called *cell division*. Life begins with one cell and progressively doubles so that 1 cell becomes 2 cells, 2 cells become 4 cells, 4 cells become 8 cells, and so on. This process gives rise to the following sequence:

$$1, 2, 4, 8, \ldots$$

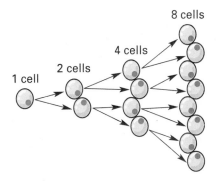

If you let y represent the number of cells present and x represent the number of cell divisions that have occurred, the sequence can be described by

$$y = 2^x$$

Does this equation describe a function?

ACTIVITY 1 **Cell Division**

Use the equation $y = 2^x$ to complete this activity.

1 Let the independent variable be x. Let the dependent variable be y. Make a table with values of x from zero through 4. Remember that $2^0 = 1$.

2 What are the restrictions on the independent variable?

3 What are the restrictions on the dependent variable?

4 Draw coordinate axes. Plot the points from your table. Connect the points with a smooth curve.

5 Explain why the equation $y = 2^x$ describes a function.

Exponential Growth

The equation $y = 2^x$ is an example of an **exponential function**. You can recognize an exponential function by observing that the independent variable is an exponent. Any function in the form $y = a^x$ where a is greater than 1 models **exponential growth**. Compare the graph of $y = 2^x$ with the graph of $y = 4^x$. As x gets larger, the graph of $y = 4^x$ rises faster.

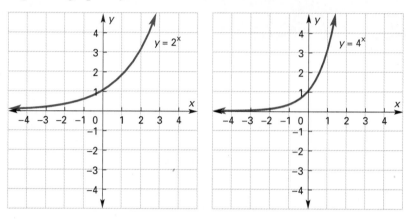

EXAMPLE 1 Increasing Bacteria

At the beginning of an experiment, a laboratory culture dish contains 500 bacteria. The number of bacteria increases by 50% each hour. After h hours, the number of bacteria (B) present is given by the formula:

$$B = 500(1.5)^h$$

1. What are the restrictions on the domain and range of the function?

2. Graph the function.

3. Estimate the number of bacteria in the culture dish after three-and-a-half hours.

SOLUTION

1. The domain includes zero and all the positive numbers. The range includes zero and all the positive numbers.

2. In this problem, the independent variable is h, and the dependent variable is B. Make a table using whole numbers as replacements for h. (Use the $\boxed{Y^x}$ key on your calculator.)

h	B
0	500
1	750
2	1125
3	1688
4	2531

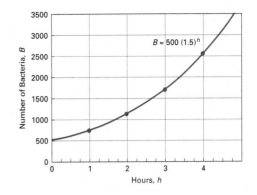

The numbers representing B are rounded to the nearest whole number. There are no fractional bacteria.

3. To find the number of bacteria after 3.5 hours, substitute 3.5 for h, and solve for B.

$$B = 500(1.5)^{3.5}$$

The exponent 3.5 is between 3 and 4. The answer should lie between 1688 and 2531. If you use the exponent key on your calculator, the result is 2066.75697. This is rounded down to about 2066 bacteria after three-and-a-half hours. Use the graph to check this approximation. Does the answer make sense?

Ongoing Assessment

Use your calculator to estimate the number of bacteria in the culture after three-and-three-quarters hours. Use the graph to check your estimate.

Exponential Decay

Some carbon atoms are radioactive. The radioactive carbon atom, carbon-14, has a half-life of about 5700 years. The half-life is the time it takes for one-half the radioactive atoms to decay.

Suppose that 22,800 years ago, a tree died. At the time of its death the tree contained 50 grams of carbon-14. After one half-life of time (5700 years) passes, the remains of the tree contain half the original amount of carbon-14 (25 grams). After two half-lives (11,400 years), the tree's remains contain one fourth the original amount of carbon-14 (12.5 grams). After three half-lives, one-eighth remains, and so on.

The following sequence describes this carbon-14 decay process:

$$1, \frac{1}{2}, \frac{1}{4}, \frac{1}{8}, \dots$$

The equation $y = \left(\frac{1}{2}\right)^x$ models this sequence. The independent variable, x, represents the number of half-lives. Does this equation describe a function?

ACTIVITY 2   Radioactive Decay

Use the equation $y = \left(\frac{1}{2}\right)^x$ to complete this Activity.

1 Let the independent variable be x. Let the dependent variable be y. Make a table for the first four values of x. Remember that $\left(\frac{1}{2}\right)^0 = 1$.

2 What are the restrictions on the independent variable?

3 What are the restrictions on the dependent variable?

4 Plot the points in your table.

5 Connect the points with a smooth curve.

6 Explain why the equation $y = \left(\frac{1}{2}\right)^x$ describes a function.

Any function in the form $y = a^x$ where a is between zero and 1, is called an **exponential decay function**. Describe the effect on the graph $y = a^x$ as x increases in value.

Critical Thinking   Suppose the decay function is multiplied by a constant. How is the graph of the parent function affected?

EXAMPLE 2   Population Change

The population of a city is 100,000. The population is declining by 5% each year. What is the population after 3 years?

SOLUTION

Each year the population is 95% of the year before. The decay function that models this situation is

$$y = 100{,}000(0.95)^x$$

Substitute 3 for x and solve for y. If you use your calculator, you will find that the population is 85,738 after three years.

Critical Thinking   Explain how to use negative exponents to make the exponential decay function look like the exponential growth function.

LESSON ASSESSMENT

1 Explain how to tell if an exponential function is a growth function.

2 Explain how to tell if an exponential function is a decay function.

3 Explain why exponential growth and decay functions have restrictions on their domain and range.

4 Explain why increasing a number by a% is the same as taking $(100 + a)$% of the original number.

5 Explain why decreasing a number by b% is the same as taking $(100 - b)$% of the original number.

Practice and Problem Solving

Sketch the graph of each function. Describe the graph.

6. $y = 3^x$ **7.** $y = 5(2)^x$ **8.** $y = 8\left(\dfrac{1}{2}\right)^x$

9. $y = 2^x + 1$ **10.** $y = 2^x - 1$ **11.** $y = \left(\dfrac{1}{2}\right)^x + 1$

12. What effect does adding or subtracting a constant to the parent exponential function have on its graph? Give an example.

13. What can you do to the parent exponential function to reflect its graph over the x-axis? Give an example.

14. Describe the equation that models this sequence:

$$1, 3, 9, 27, \ldots$$

Write the function that models each situation. Solve each problem.

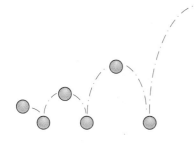

15. Suppose a ball bounces to one-half of its previous height on each bounce. Make a table of values for the first four bounces. Draw the graph.

16. The population of a city is 50,000. It is increasing at an annual rate of 3%. What is the population at the end of 4 years?

17. An investment compounds annually at 6% per year is kept for 5 years. Write the exponential function that models the amount in the investment after each year. If the initial amount invested was $1000, how much is in the account after 5 years?

18. Suppose you start with a penny and someone doubles the amount you get every day. How much will you receive on the tenth day?

19. The value of a car depreciates by 8% each year. If a new car is purchased for $20,000, what is the car worth after 4 years?

Mixed Review

20. A light-year (5.88×10^{12} miles) is the distance light travels in one year. The star Vega is 23 light-years from Earth. Write the distance (in miles) from Vega to Earth in scientific notation.

Write an equation and solve.
21. You make 6% commission on all of your sales. If your commission is $250, how much are your sales?

22. A spherical tank has a radius of 4 meters. Find the volume of the tank.

23. If $f(x) = 5x + 9$, what is $f(-3)$?

24. If $g(r) = r^2 - r$, what is $g(-5)$?

25. Find the equation of the line passing through the points $A(1,3)$ and $B(-2,5)$.

LESSON 6.6 THE RECIPROCAL FUNCTION

In Lesson 4.2 you used the reciprocal of a number to solve equations where the coefficient of the variable is a rational number.

For example,

$$\frac{3}{4}x = 12 \qquad \text{Given}$$

$$\frac{4}{3} \cdot \frac{3}{4}x = \frac{4}{3} \cdot 12 \qquad \text{Multiplication Property of Equality}$$

$$x = 16 \qquad \frac{4}{3} \cdot \frac{3}{4} = 1 \text{ and } \frac{4}{3} \cdot 12 = 16$$

The Multiplication Property of Equality is effective in solving this equation because $\frac{4}{3}$ is the reciprocal of $\frac{3}{4}$. Remember, the product of a number and its reciprocal is always one. Multiplying by the reciprocal of the coefficient of x isolates x on one side of the equation.

You can extend the idea of a reciprocal to functions. The equation that describes the **reciprocal function** is

$$y = \frac{1}{x} \ (x \text{ cannot equal zero})$$

Notice that in a reciprocal function the independent variable is in the denominator of the function.

ACTIVITY 1   Graphing Part of the Function

1 Complete the table for positive values of x.

x	$\dfrac{1}{x}$
4	?
2	?
1	?
$\dfrac{1}{2}$	?
$\dfrac{1}{4}$	?

2 Plot the points for $4 = \dfrac{1}{x}$ and connect the points.

3 What are the restrictions on the domain and range?

4 What happens to y when x is very large?

5 What happens to y when x is very small?

6 What happens if you try to substitute zero for x?

ACTIVITY 2   The Rest of the Story

1 Repeat the same steps used in Activity 1. But this time, use the opposite of each value in the domain.

2 Put the two graphs together. Describe what you see. Why is this a representation of a function?

The graph of the function $y = \dfrac{1}{x}$ is called a *hyperbola.* The graph is made up of two branches, one in the first quadrant and the other in the third quadrant.

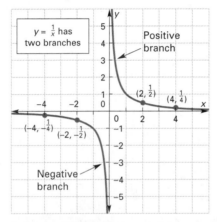

Critical Thinking   Explain why the branches of the graph of a hyperbola do not touch each other. Explain why the branches of the graph get closer and closer to the axes but never touch.

Inverse Variation

The reciprocal function is an example of a relation called inverse variation. In **inverse variation**, as one variable gets larger, the other variable gets smaller.

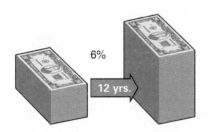

A formula called the Rule of 72 is an inverse variation used to approximate how fast money will double when it is invested at a given compound interest rate. The number of years (y) to double an investment is equal to 72 divided by the annual interest rate (r) expressed as a percent.

$$y = \frac{72}{r}$$

If you invest at a rate of interest of 6%, it will take about $\frac{72}{6}$, or 12 years, for your money to double.

How long will it take to double your money if the interest rate is 7%?

Critical Thinking   If $y = \dfrac{1}{x}$ is the parent reciprocal function, what effect does multiplying the function by a positive constant have on its graph?

For example, light intensity follows an *inverse square law*. The intensity (I) of light that falls on an object at a given distance (d) from a light source of power (P) is given by this formula:

$$I = \frac{P}{d^2}$$

If the distance is measured in meters and the power is measured in lumens, then the intensity is measured in lumens per square meter.

EXAMPLE Brightness of Light

A photographer takes pictures for a store catalog. The item being photographed is 3 meters from a 72 lumen light source. Where should the object be placed to receive twice as much light?

SOLUTION

To find the intensity when the object is 3 meters from the light source, substitute 72 for P, 3 for d, and solve for I. Since 3^2 is 9, and 72 divided by 9 is 8, the intensity is 8 lumens per square meter.

But you need to find the distance at which the luminosity is twice as much, or 16 lumens per square meter. Replace I by 16, and solve for d.

$$16 = \frac{72}{d^2} \qquad \text{Given}$$

$$16d^2 = 72 \qquad \text{Multiplication Property}$$

$$d^2 = 4.5 \qquad \text{Division Property}$$

$$d = \pm\sqrt{4.5} \qquad \text{Solving } x^2 = c$$

The distance is about 2.1 meters. Why is the negative solution not used?

The Pencil or Vertical-Line Test

You can use a technique called the "pencil test" or "vertical-line test" to distinguish between relations that are functions and those that are not. With this test, move a pencil (oriented as a vertical line) from left to right across the graphed curve of the relation. As the pencil moves, check to see how many points of the curve it intersects at any one time. If the pencil crosses the graph of the relation one point at a time, the relation is a function.

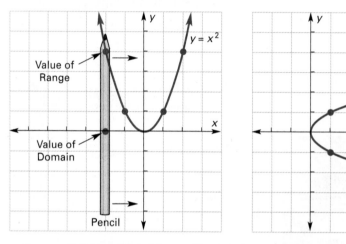

Critical Thinking Explain how the vertical-line test shows that $y = x^2$ is a function but $y = \pm\sqrt{x}$ is not a function.

LESSON ASSESSMENT

Think and Discuss

1 How can you find the reciprocal of a number?

2 Why is there no reciprocal for zero?

3 How are hyperbolas and parabolas alike? How are they different? Draw a picture to illustrate your explanation.

4 Explain the meaning of inverse variation. How is inverse variation different from direct variation?

5 Explain how the vertical-line test can be used to test for a function. Give an example.

Graph each function. Use a graphics calculator if possible.

6. $y = 2\left(\dfrac{1}{x}\right)$ **7.** $y = 5\left(\dfrac{1}{x}\right)$ **8.** $y = 10\left(\dfrac{1}{x}\right)$

9. What effect does multiplying the parent reciprocal function by a number greater than 1 have on its graph?

Graph each function. Use a graphics calculator if possible.

10. $y = -2\left(\dfrac{1}{x}\right)$ **11.** $y = -5\left(\dfrac{1}{x}\right)$ **12.** $y = -10\left(\dfrac{1}{x}\right)$

13. What effect does multiplying the parent reciprocal function by a number less than -1 have on its graph?

Graph each function. Use a graphics calculator if possible.

14. $y = \dfrac{1}{2x}$ **15.** $y = \dfrac{1}{5x}$ **16.** $y = \dfrac{1}{10x}$

17. What effect does multiplying the parent reciprocal function by a number greater than zero but less than 1 have on its graph?

In computer programming, a convenient function for rounding is the *greatest integer function*. This function, written $y = \text{INT}(x)$, is defined as follows:

A number (x) written in decimal form is rounded down to the nearest integer (y). For example,

$$\text{INT}(3.1459) = 3 \qquad \text{INT}(-3.1459) = -4$$
$$\text{INT}\left(\dfrac{3}{4}\right) = 0 \qquad \text{INT}(-1) = -1$$

18. What is the domain of the INT function?

19. What is the range of the INT function?

20. Graph the INT function for the numbers in the domain that are between -5 and 5.

21. Describe the graph of the INT function.

22. How can you tell by looking at the graph that the INT function is a function?

23. How many 32-cent stamps can you buy for $1? Explain why this situation can be modeled by an INT function.

24. Explain how to evaluate the greatest integer function using your calculator.

When taking a 1-minute timed test, you stop at 55 seconds.

25. Are you over or under the allotted time?

26. Use an integer to show how much you are over or under.

27. What is the absolute value of your miss?

Suppose you are going to tile the floor shown here.

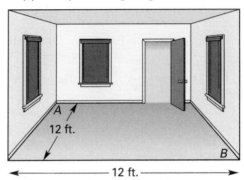

28. If you want to use tiles that are 9 inches by 9 inches, how many tiles will you need?

29. What is the distance from corner *A* to corner *B*?

Thomas needs to cut a 72-inch board into three pieces with the following specifications. One piece is five inches longer than the shortest piece and five inches shorter than the longest piece.

30. Write an equation to model the situation.

31. Find the length of the shortest piece.

32. Find the length of the longest piece.

MATH LAB

Equipment Calculator
Pizza price list showing prices of at least four different pizzas

Problem Statement

A price list for pizza usually shows the prices for pizzas of different diameters. The price increases as the diameter increases. You will examine a price list for several different size pizzas with the same ingredients to determine the relationship of the price to the diameter.

Procedure

a Make a table with columns for diameter, area, and price.

b Record the diameter and price for each pizza in the table.

c Calculate the area of each size of pizza. Record the area in the table.

d Use your data to make a graph. Use the *x*-axis for the diameter. Use the *y*-axis for the price.

e Now make a second graph using the *x*-axis for the diameter and the *y*-axis for the area. Compare the two graphs you have drawn.

f Now make a third graph using the *x*-axis for area and the *y*-axis for price. What do you notice?

g Is the graph of pizza area versus price *linear?* If so, what is the slope of the graphed line? What are the units of this slope? What is the significance of this slope?

h Explain why pizza shops would lose money if their prices were proportional to the diameter of their pizza instead of the area.

Equipment   Calculator
Checkerboard
1 pound of kitty litter

Problem Statement

You will examine how organisms such as bacteria increase in number if they double at regular intervals. You will do this by recording how quantities of objects increase as they are doubled when you move from one checkerboard square to the next. The equation that describes this doubling process is

$$N = 2^{(x-1)}$$

N is the number of objects present, and x is the number of doubling intervals. You will use grains of cat litter as objects, and each square of a checkerboard will represent a doubling interval.

Procedure

a Place one grain of cat litter on square 1, two grains on square 2, four grains on square 4, and so on until you reach square 8.

b For squares 1 through 8, make a table showing the number of each square and the corresponding number of grains of cat litter on that square.

c Use your table to draw a graph. Use the x-axis for the number of the square and the y-axis for the number of grains. Is the graph linear? Why or why not?

d Use the formula $N = 2^{(x-1)}$ and your calculator to determine the number of grains on each of the next 8 squares. Continue your table.

e Make a graph of the new data. Is this graph similar to the one you have already drawn?

f Find the sum of the grains on the first 7 squares. How does the sum compare to the number of grains on square 8?

g Find the sum of the grains on the first 15 squares. How does the sum compare to the number of grains on square 16?

h What pattern do you see in the last two steps? Use this pattern to determine how many grains will be placed on square 64.

i Based on the pattern you have discovered, decide which is the better deal:

1. You can have ten million dollars today; or

2. You can start with $1 and double the amount every day for a month.

Activity 3: Height and Time of Bouncing Ball

Equipment   Calculator
Tape measure
Timer
Rubber ball

Problem Statement

When a ball is dropped to the floor from a given height, the time it takes to stop bouncing depends on the height from which it is dropped.

Procedure

a Locate an area where there is a hard, smooth floor. Drop the ball onto the floor from a height of 6 feet. Measure the length of time that the ball bounces. Record the drop height and time.

b Repeat the drop three more times from the same height. Average the four trials and record the average in a table.

c Repeat steps **a** and **b** using heights of 5 feet, 4 feet, 3 feet, 2 feet, and 1 foot.

d Graph your data. Use the *x*-axis for the average drop height and the *y*-axis for the average time the ball bounces.

e Is the relationship between drop height and time linear or nonlinear?

MATH APPLICATIONS

The applications that follow are like the ones you will encounter in many workplaces. Use the mathematics you have learned in this chapter to solve the problems.

Wherever possible, use your calculator to solve the problems that require numerical answers.

You have a 12-foot extension ladder. This ladder, when fully extended, is about 12 feet long. When retracted, it is about 6 feet long. Suppose you place the base of the ladder 2.5 feet away from a wall.

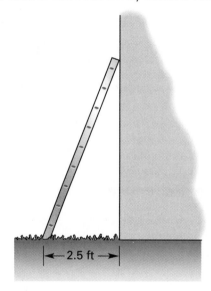

2.5 ft

1 Use the Pythagorean formula to obtain an equation for the reaching height of the ladder for extensions from 6 feet to 12 feet.

2 Is this equation linear or nonlinear?

3 Make a table of reaching heights for several values of the ladder extension between 6 feet and 12 feet. Draw a graph of the reaching heights versus ladder extensions. Does your graph confirm your answer to Exercise 2? Explain why or why not.

You can predict the balance of your savings account each month. This is easy with a scientific calculator using the formula

$$B = P(1 + i)^n$$

B is the balance after n months.

P is the starting balance (assuming you make no deposits or withdrawals).

i is the percent interest paid each month, as a decimal value.

n is the number of months.

4 Is the formula a linear or nonlinear equation? Explain your answer.

5 For an annual interest rate of 6%, the monthly interest rate would be $\frac{1}{12}$ of 6%, or 0.5%. For a starting balance of $1000, make a table showing the growth of this deposit. Select several different numbers of months between zero and 120 months (that is, a span of 10 years). (Hint: Be sure to use the decimal value for the interest rate.)

6 Add another column to your table, computing the balances for the same values of n for an annual interest rate of 7%.

7 Draw a graph of the balances you calculated. Connect the sets of points from each percentage rate with a smooth curve. Does the appearance of the graph agree with your answer to Exercise 4?

When you travel at a constant speed of 0.75 mile per minute (that is, 45 miles per hour), the total mileage D you travel during t minutes can be computed by the formula

$$D = 0.75t$$

On the other hand, if you accelerate and change your speed by 0.0167 mile per minute (that is, one mile per hour) each minute, the distance D you can travel in t minutes can be computed by the formula

$$D = \frac{1}{2}(0.0167)t^2$$

8 Make a table of values for D for each case above, (**a**) with a constant speed of 0.75 mile per minute, and (**b**) accelerating 0.0167 mile per minute each minute. Use values of t from 0 to 120 minutes.

9 Draw a graph of the values in your table and connect each set of points with a smooth line or curve. Label each line (or curve) as either "constant speed" or "constant acceleration."

10 Suppose you travel for 60 minutes with each of the methods described above. With which method would you be farther from your starting position? If you continued on for another 120 minutes, with which method would you be farther along?

11 Describe a method to find at what time t both methods would have you at the same distance from the starting point.

12 On a certain golf course, the traditional play is to take two strokes to go around a small lake. Each stroke requires a drive of about 250 yards. Determine how long a drive is needed to cross the lake in one stroke.

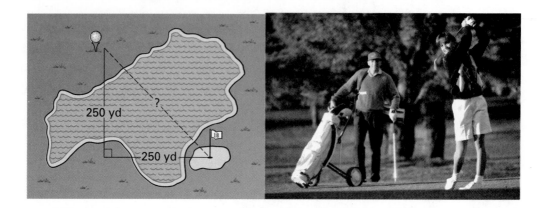

The water content of soil affects its weight. The results of certain soil tests produce the results shown below as a graph (*called a compaction curve*).

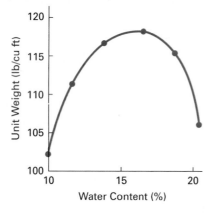

13 Examine the curve that is drawn through the test data. Is the trend in the data linear or nonlinear?

14 The most important feature of the curve is the maximum (that is, the high point of the peak). The maximum weight shown on the graph is called the *maximum dry density,* or the M.D.D. The percent water content at which this maximum occurs is called the *optimum moisture,* or O.M., of the soil. Approximately what are the M.D.D and the O.M. of this soil sample?

One of the basic tests performed on soils is the sieve test. The soil sample is allowed to settle through a stack of sieves that are of progressively finer and finer mesh. The amount of soil that remains in each sieve is measured. The percent of the soil sample that passes through each sieve is computed. The results from a typical soil sample are tabulated below.

Sieve Size	Soil Particle Size (mm)	% Finer
3 in.	76.2	92%
2 in.	50.8	90%
1 in.	25.4	89%
1/2 in.	12.7	82%
No. 4	4.76	73%
No. 10	2.00	66%
No. 20	0.84	56%
No. 40	0.42	41%
No. 60	0.25	25%
No. 100	0.15	12%
No. 200	0.07	4%

15 Draw a graph of the percentage finer versus the soil particle size for percentages less than 75%. Sketch in a curve that seems to fit the points on your graph.

16 Does the graph seem to be linear or nonlinear?

17 Using your sketched curve, estimate the particle sizes that correspond to percentages of 10% and 60%. (The particle size corresponding to 10% is called the *effective size*.)

18 Compute the ratio of the two particle sizes determined above: the size for 60% to the size for 10%. (This ratio is called the *uniformity coefficient*.)

Lumber mill operators estimate usable log volume from the formula below, which allows for saw kerf, slab, and edging deductions:

$$V = 0.0655L\,(1 - A)\,(D - S)^2$$

V is the usable volume of a log in board feet.

L is the length of the log in feet.

A is the decimal value of the percent deduction for saw kerf.

D is the log diameter in inches.

S is the slab and edging deductions in inches.

19 Rewrite the equation above for logs that are 32 feet long, cut with a 10% saw kerf, and have 3 inches of slab and edging deductions.

20 Make a table of the usable board feet of lumber that can be obtained under these conditions for logs with diameters of 10 in., 15 in., 20 in., 30 in., and 40 in. (Round your answers to the nearest 10 board feet.)

21 Draw a graph of this data.

22 What is the meaning of the graph as it approaches the x-axis? At what diameter does the usable number of board feet equal zero, based on this equation?

BUSINESS & MARKETING

The graph below shows the future value of a savings account earning 6% annual interest when $100 monthly deposits are made.

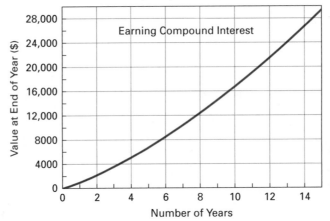

23 Does the graph show a linear relationship between the number of years and the savings account balance?

24 Compute the slope of the line that is formed by using the origin and the balance at 10 years as the endpoints.

25 Compare this slope with the slope of the line representing $100 per month deposits. Explain this comparison. Use the graph to estimate the amount of interest accumulated in the account after 10 years.

26 Suppose you determined the slope of the line formed by using the origin and the balance at 15 years. Would this slope be greater or less than the slope you calculated in Exercise 24? Explain your answer.

The graph below illustrates the changing labor costs in the American automotive industry for the years from 1960 through 1984.

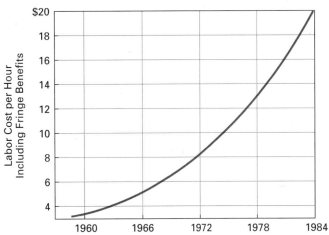

27 Is this a graph of a linear trend or a nonlinear trend?

28 The shape of the curve most resembles which relationship?

a. $y = x^2$ **b.** $y = \sqrt{x}$

c. $y = \dfrac{1}{x}$ **d.** $y = |x|$

29 Compute the slope of the small section of the graph for the period of the 1960s, and then for the 1970s. Round your answer to the nearest $0.01 per hour. Did you get about the same values? Explain.

The "Rule of 72" is a simple way to observe the effect of compound interest. With this rule you can estimate the number of years needed to double an investment:

$$Y = \frac{72}{i}$$

Y is the years needed to double an investment.

i is the annual percentage at which the account earns interest, such as 8.5 for 8.5%.

30 The equation above most closely resembles which of the following nonlinear forms:

a. $y = x^2$ **b.** $y = \sqrt{x}$

c. $y = \dfrac{1}{x}$ **d.** $y = |x|$

31 Make a table of values for Y for interest rates of 3%, 4%, 5%, 6%, 7%, and 8%.

32 Draw a graph of the data in your table and connect the points with a smooth curve.

33 Is the slope of your curve steeper for smaller interest rates or for larger interest rates? Explain what this means.

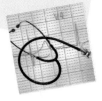

HEALTH OCCUPATIONS

When the count of red blood cells increases, there is more friction between the layers of cells. This friction determines the thickness, or viscosity, of the blood. A common measure of the blood—cell count—is the *hematocrit*, the percentage of the blood by volume that is cells. A graph of the effect of various hematocrit values is shown below. (Note that the viscosity of water is shown as equal to 1 on the graph, for comparison with whole blood and with plasma.)

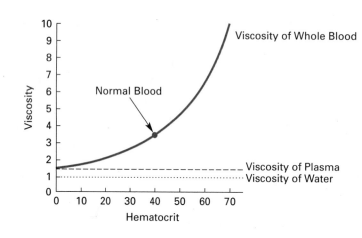

34 Which of the curves in the graph appears to be nonlinear?

35 Which of the following relationships does the nonlinear graph appear to resemble:

a. $y = x^2$ **b.** $y = \sqrt{x}$

c. $y = \dfrac{1}{x}$ **d.** $y = |x|$

36 Investigate the nonlinear curve for low values of hematocrit as well as for higher values. In which case does a small change in the hematocrit have more effect on the viscosity?

The vapor pressure of water is an important factor in inhalation therapy. A table of the vapor pressures of water (in units of millimeters of mercury) for various temperatures is shown below.

Temperature (°C)	Pressure (mm Hg)
0	4.6
10	9.1
20	17.4
30	31.5
40	54.9
50	92.0
60	148.9
70	233.3
80	354.9
90	525.5
100	760.0

37 Draw a graph of the vapor pressure of water for the temperatures listed in the table above.

38 Which of the following relationships does the nonlinear graph resemble:

a. $y = x^2$ **b.** $y = \sqrt{x}$

c. $y = \dfrac{1}{x}$ **d.** $y = |x|$

Doctors will frequently order IV (intravenous) fluids for patients needing additional body fluids. The hospital staff adjusts the IV drip rate so that the patient receives the prescribed amount of fluid in the prescribed amount of time. The drip rate can be determined using the formula

$$D = \frac{V}{4T}$$

D is the drip rate, in drops per minute.

V is the prescribed volume of fluid, in cubic centimeters (cc).

T is the prescribed amount of time for the fluids to be given to the patient, in hours.

39 A common volume for IV fluids is 1000 cc. Rewrite the equation that would be used to determine the drip rate for this volume, for various times T.

40 A burn victim may require such a volume to be given rapidly—perhaps in 3 to 4 hours. For a more stable patient, you may only wish to keep the vein open by a very slow flow—a total time span as long as 24 hours. Make a table of the drip rates indicated by your equation for times ranging from 3 hours to 24 hours.

41 Draw a graph of this equation using your table of data from above.

INDUSTRIAL TECHNOLOGY

Police officers frequently investigate the scenes of automobile accidents. By measuring the length of the skid marks, they can make a reasonable estimate of a car's speed before the skid began. Below is a table of average skid distances for an automobile with good tires on dry pavement.

Speed (mph)	Skid Distance (ft)
20	19
30	38
40	70
50	110
60	156
70	215
80	276

42 A graph of these data might make it easier to translate skid marks to estimated speeds. Choose a scale for your axes and make a graph of the data.

43 Draw a line or a curved line that seems to match the data. Should your line (or curved line) pass through the origin? Explain. Which of the following does the graph resemble?

a. $y = x^2$　　　　**b.** $y = \sqrt{x}$　　　　**c.** $y = \dfrac{1}{x}$

d. $y = |x|$　　　　**e.** $y = x$

44 Suppose you measured a skid mark of 90 feet. Use your graph to estimate the speed the car was traveling at the time the skid began.

While adjusting the spark advance on a high-performance engine, you monitor the exhaust valve temperature. Below is a table of the data you collect.

Spark Advance (degrees)	Exhaust Valve Temperature (°F)
10	1350
20	1280
30	1255
40	1255
50	1300

45 Choose an appropriate scale and graph the data.

46 Is this graph linear or nonlinear? Draw a straight or curved line that seems to match the data.

47 If you want to set the spark advance to obtain the lowest temperature, what setting would you choose? Show this on your graph.

48 A surveyor is establishing lot lines for a housing development. To find the distance across a lake between two points *A* and *B*, she measures the two perpendicular distances shown. Determine the distance across the lake to the nearest foot.

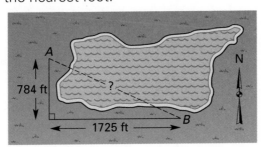

Below are two graphs that show a relative comparison of the speed versus torque (or turning power) for a direct-current (dc) motor and an alternating-current (ac) motor.

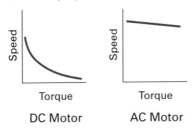

49 Identify whether each graph appears linear or nonlinear.

50 For the dc motor, describe what happens to the motor speed when the torque on the motor is very small.

51 How does this trend compare to the ac motor?

52 Select one of the graphs above that is nonlinear. Which of the following relationships does the graph appear to resemble?

    a. $y = x^2$ **b.** $y = \sqrt{x}$

    c. $y = \dfrac{1}{x}$ **d.** $y = |x|$

The thermistor, the RTD, and the thermocouple are three common electronic measuring devices. A graph of the measurement sensitivity (either voltage or resistance) is shown for each of these types of devices.

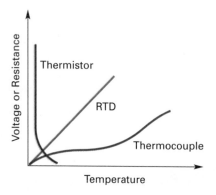

53 Identify which of the three curves are nonlinear and which are linear, or very nearly linear.

54 The thermistor curve is very different from the RTD and thermocouple curves. This curve most closely resembles which of these relationships:

a. $y = x^2$ **b.** $y = \sqrt{x}$

c. $y = \dfrac{1}{x}$ **d.** $y = |x|$

55 What advantage would there be in using a thermistor for temperature measurements?

A technical handbook for temperature measuring equipment shows the following graph of response times for various types of thermocouple constructions.

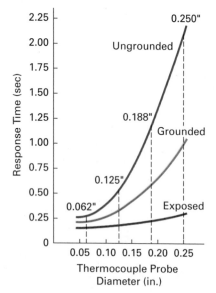

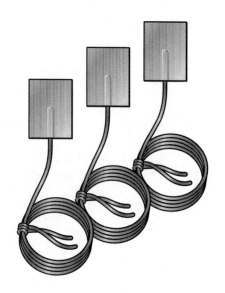

56 Identify the three types of thermocouple probes shown by the three curves in the graph.

57 Based on this graph, which type of thermocouple has a response time that is most sensitive to the diameter of the probe? Explain your answer.

58 If you needed a probe with a short response time (less than half a second) and a diameter of about 0.05 inch, would one type of probe perform significantly better than another? What if you were using probes with diameters of 0.25 inch?

59 Could you use a linear equation to approximate any of the regions of any of these curves? If so, describe those regions.

With the rising utility of superconductors, measurement of very low temperatures has become critical. The temperature scale of Fahrenheit is no longer convenient, and so the Kelvin temperature scale is used. In the Kelvin scale, 0°K is called *absolute zero* (the lowest possible temperature), and the freezing point of water is at 273°K. Some superconducting devices operate very close to 0°K. To measure these low temperatures requires special equipment. Below is a voltage response curve for one type of sensor.

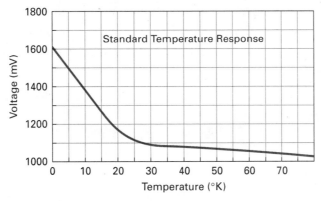

60 Identify two ranges of temperatures in the graph for which the response curve is very nearly linear.

61 For each of the ranges above, determine the approximate slope of the graph in the region. (Round to the nearest 1 mV per °K.)

62 What advantage is it for the slope to be steeper in the region close to 0°K?

The force of drag on a car is an important design consideration. By measuring how much the speed drops during a 10-second interval of coasting, an estimate of the drag force on the car can be found. The force of drag is related to the change in speed divided by the amount of time. The table lists data from such a test.

Speed(km/hr)	Drag Force (kg force)
20	29
30	30
40	34
50	38
60	43
70	50
80	59
90	68
100	75
110	86

63 Draw a graph of the data listed above and draw a smooth curve through the plotted points.

64 Does the relationship appear to be linear or nonlinear?

65 Based on your graph, can you predict what the drag force would be at a speed of zero? Explain.

The reference manuals for many automobiles specify how hard you should tighten bolts in various places on the car. This is done by specifying the "torque" to apply to the bolt. The torque is equal to the force multiplied by the length of the lever arm.

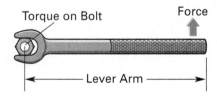

Torque on Bolt   Force

Lever Arm

You can compute the force needed when using wrenches of different lengths by rearranging the relationship to obtain the expression below.

$$Torque = Force \bullet Lever\ Arm$$

Torque is measured in foot-pounds.

Lever Arm is measured in feet.

Force is measured in pounds.

66 Rewrite the equation to show *Force* as a function of *Lever Arm*. For a torque of 45 foot-pounds, make a table and graph of the dependent variable when the independent variable ranges in value between 0.5 and 3.

67 Is this relationship linear or nonlinear?

68 What is the difference in required force if you change from a small wrench that is 6 inches long to one that is 12 inches long? What difference is there when going from a wrench 2 feet long to a wrench 3 feet long?

Ethylene glycol is commonly used as an antifreeze. When added to the water in a car radiator it lowers the freezing point. The table below shows the freezing point for various concentrations of this antifreeze.

Concentration (%)	Freezing Point (°C)
0	0
9.2	−3.6
18.3	−7.9
28	−14
37.8	−22.3
47.8	−33.8
58.1	−49.3

69 Draw a graph of the freezing points for the various concentrations listed above. Connect the points with a smooth curve or line. Does your curve suggest a linear relationship or a nonlinear one?

70 Use your graph to estimate what concentration would provide protection down to − 40°C.

71 Someone suggests simply using pure antifreeze, with a concentration of 100%, reasoning that this would provide the lowest temperature protection. Can you use your graph to estimate what freezing point protection this would give you? Explain your answer. (Hint: The freezing point of 100% ethylene glycol is actually about −17°C.)

A pile driver is a large mass used as a hammer to drive pilings into soft earth to support a building. The hammer is lifted to a certain height and allowed to drop freely, striking the piling and driving it into the ground. The velocity with which the pile driver strikes the piling is related to the height from which it is dropped by the equation

$$v^2 = 64.4h$$

v is the final velocity of the hammer, in feet per second.

h is the distance that the hammer falls, in feet.

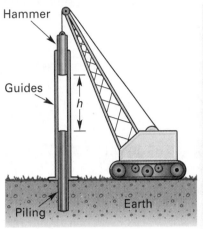

72 Solve the equation above for the variable *v*.

73 Draw a graph of your equation for various values of *h* from zero to 25 feet.

74 This is clearly a nonlinear equation because of the radical. Can any portion of your graph be approximated by a linear relationship? If so, what portion?

You are designing an electric circuit that might have many voltage cells (batteries). Each cell has its own resistance of 5 ohms in addition to the 16-ohm resistance of the rest of the circuit. Each cell has a voltage of 1.4 volts. Below is an equation that predicts the current for such a circuit, depending on the number of cells used.

$$i = \frac{nV}{R + nr}$$

i is the current in the circuit, in amperes.

n is the number of cells used.

V is the voltage supplied by each cell, in volts.

r is the internal resistance of each cell, in ohms.

R is the resistance of the rest of the circuit, in ohms.

75 Substitute the known values into the equation above to obtain an equation for *i* dependent on the number of cells used in the circuit.

76 Make a table of the current predicted by the equation for values of *n* from zero to 10 cells.

77 Draw a graph of the values in your table. Explain why it would not make sense to connect the points in this graph.

Rooftops of houses are often built in the shape of a pyramid with a square base. The *lateral area* of a pyramid is the total surface area of the pyramid minus the area of the base. The lateral area of this pyramid can be expressed by the formula

$$L = 2s \sqrt{0.25s^2 + h^2}$$

L is the lateral area of the pyramid-shaped surface.

s is the width of the base.

h is the height of the peak.

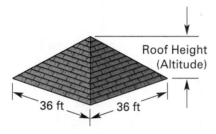

Roof Height
(Altitude)

36 ft   36 ft

78 For a house that has each base width equal to 36 feet, as shown in the figure, rewrite the formula to obtain an equation for *L* in terms of the height of the roof peak, *h*.

79 Suppose each bundle of shingles covers 25 square feet of roof. Rewrite the equation to allow you to compute the number of bundles, *n*, needed to cover any pyramid-shaped room with a square base.

80 Make a table of the number of bundles of shingles needed for each height.

81 Draw a graph of your data. Sketch a smooth curve through your points. Compare your graph to the graph of $y = \sqrt{x}$. How are they alike and how are they different?

CHAPTER 6 ASSESSMENT

Skills

If $f(x) = 3x - 5$, find

1. $f(2)$ **2.** $f(-4)$ **3.** $f\left(\dfrac{5}{3}\right)$

If $f(x) = -3x^2 + 2$, find

4. $f(3)$ **5.** $f(-1)$ **6.** $f\left(\dfrac{1}{2}\right)$

Graph on a coordinate system:

7. $y = 4|x|$ **8.** $y = |x| - 4$ **9.** $y = \dfrac{1}{4}|x|$

10. $y = x^2 + 4$ **11.** $y = -2x^2 - 1$ **12.** $y = \dfrac{1}{2}x^2 + 3$

13. $y = 2^x + 3$ **14.** $y = \left(\dfrac{1}{3}\right)^x$ **15.** $y = |x + 3|$

16. Give the domain and range of each function.

a. $y = x^2$ **b.** $y = \sqrt{x}$

Applications

17. The width of a rectangular room is 12 feet, and the length is 16 feet. What is the length of the diagonal of the room?

18. A Navy plane is 2000 feet above and at a 5,000 foot horizontal distance from the end of an aircraft carrier deck. How far from the end of the deck is the plane?

19. A rectangular carton is 45 inches wide and 60 inches long. Can a skateboard that is 70 inches long fit in the box? Why or why not?

20. What is the width of a rectangle that has an area of 93 square feet and a length of 10 feet?

Math Lab

21. Which pizza is the better bargain?

    a. a10 inch pizza for $5.00
    b. a 12 inch pizza for $7.00

22. Which gives the larger ending balance?

    a. an initial amount of $100,000 that compounds at an annual interest rate of 10% for 30 years
    b. an initial amount of $0.01 that doubles every year for 30 years.

23. You drop a ball from a height of one foot and observe that it takes 11 seconds for the ball to stop bouncing. You drop the same ball from 2 feet and observe the ball bounces for 25 seconds. The next drop will be from 3 feet. How long will the ball bounce?

    a. 39 seconds
    b. less than 39 seconds
    c. more than 39 seconds

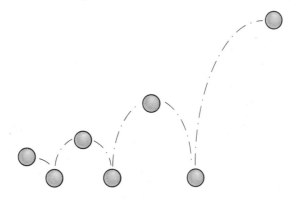

Tables of Conversion Factors

To convert from meters to inches, for example, find the row labeled "1 meter" and the column labeled "in." The conversion factor is 39.37. Thus, 1 meter = 39.37 in.

LENGTH

	cm	m	km	in.	ft	yd	mi
1 centimeter	1	0.01	10^{-5}	0.3937	3.281×10^{-2}	1.094×10^{-2}	6.214×10^{-6}
1 meter	100	1	10^{-3}	39.37	3.281	1.094	6.214×10^{-4}
1 kilometer	10^5	1000	1	3.937×10^4	3281	1094	0.6214
1 inch	2.54	0.0254	2.54×10^{-5}	1	0.0833	0.0278	1.578×10^{-5}
1 foot	30.48	0.3.48	3.048×10^{-4}	12	1	0.3333	1.894×10^{-4}
1 yard	91.44	0.9144	9.144×10^{-4}	36	3	1	5.682×10^{-4}
1 mile	1.6093×10^5	1609.3	1.6093	6.336×10^4	5280	1760	1

AREA

	cm²	m²	in.²	ft²	acre	mi²	ha
1 square centimeter	1	10^{-4}	0.1550	1.076×10^{-3}	2.471×10^{-8}	3.861×10^{-11}	10^{-8}
1 square meter	10^4	1	1550	10.76	2.471×10^4	3.861×10^{-7}	10^{-4}
1 square inch	6.452	6.452×10^{-4}	1	6.944×10^{-3}	1.594×10^{-7}	2.491×10^{-10}	6.452×10^{-8}
1 square foot	929.0	0.09290	144	1	2.296×10^{-5}	3.587×10^{-8}	9.29×10^{-6}
1 acre	4.047×10^7	4047	6.273×10^6	43,560	640	1.563×10^{-3}	0.4047
1 square mile	2.590×10^{10}	2.590×10^6	4.007×10^9	2.788×10^7	640	1	259
1 hectare	10^8	10^4	1.55×10^7	1.076×10^5	2.471	3.861×10^{-3}	1

VOLUME (CAPACITY)

	cm³	m³	in.³	ft³	L	oz	gal
1 cubic centimeter	1	10^{-6}	0.06102	3.531×10^{-5}	1.000×10^{-3}	0.03381	2.642×10^{-4}
1 cubic meter	10^6	1	6.102×10^4	35.31	1000	3.381×10^4	264.2
1 cubic inch	16.39	1.639×10^{-5}	1	5.787×10^{-4}	0.01639	0.5541	4.329×10^{-3}
1 cubic foot	2.832×10^4	0.02832	1728	1	28.32	957.5	7.480
1 liter	1000	1.000×10^{-3}	61.03	0.03532	1	33.81	0.2642
1 ounce	29.57	2.957×10^{-5}	1.805	1.044×10^{-3}	0.02957	1	7.813×10^{-3}
1 gallon	3785	3.785×10^{-3}	231	0.1337	3.785	128	1

1 gallon = 4 quarts (qt) = 8 pints (pt) = 16 cups (c)
1 cup (c) = 8 ounces (oz) = 16 tablespoons (tbsp) = 48 teaspoons (tsp)

MASS/WEIGHT

	g	kg	oz	lb	ton (short)	ton (metric)
1 gram	1	10^{-3}	0.03527	2.205×10^{-3}	1.102×10^{-6}	10^{-6}
1 kilogram	10^3	1	35.27	2.205	1.102×10^{-3}	10^{-3}
1 ounce	28.35	0.02835	1	0.0625	3.125×10^{-5}	2.835×10^{-5}
1 pound	453.6	0.4536	16	1	0.0005	4.536×10^{-4}
1 ton (short)	9.072×10^5	907.2	3.2×10^4	2000	1	0.9072
1 ton (metric)	10^6	10^3	3.527×10^4	2.205×10^3	1.102	1

ANGLE

	min (')	deg(°)	rad	rev
1 minute	1	0.01667	2.909×10^{-4}	4.630×10^{-5}
1 degree	60	1	0.01745	2.778×10^{-3}
1 radian	3438	57.30	1	0.1592
1 revolution	2.16×10^4	360	6.283	1

1 gallon = 4 quarts (qt) = 8 pints (pt) = 16 cups (c)
1 cup (c) = 8 ounces (oz) = 16 tablespoons (tbsp) = 48 teaspoons (tsp)

TIME

	s	min	hr	d*	y*
1 second	1	0.01667	2.788×10^{-4}	1.157×10^{-5}	3.169×10^{-8}
1 minute	60	1	0.01667	6.944×10^{-4}	1.901×10^{-6}
1 hour	3600	60	1	0.04167	1.141×10^{-4}
1 day*	8.640×10^4	1440	24	1	2.738×10^{-3}
1 year*	3.156×10^7	5.259×10^5	8766	365.3	1

Glossary

Absolute value The distance of a number from zero on a number line.

Absolute value of a vector The length of a vector, its numerical value. Also called the magnitude.

Addition property of equality When you add the same quantity to both sides of an equation, the equation stays equal and balanced.

Additive identity Zero is called the additive identity because when you add zero the value of the original quantity remains unchanged. Example: $5 + 0 = 5$

Algebraic expression A mathematical sentence involving at least one variable.

Area The amount of space inside a flat shape, such as a triangle or a circle.

Associative property of addition When you add multiple terms, you can change the way they are grouped without changing the answer.
Example: $5 + (3 + 17) = (5 + 3) + 17$

Associative property of multiplication When you multiply several terms, you can change the way they are grouped without changing the answer.
Example: $5 \times (3 \times 17) = (5 \times 3) \times 17$

Axis A number line used to locate points on a graph. Locating points on a plane requires two axes.

Balancing equations Whenever you perform an operation on one side of an equation, you must perform the same operation to the other side in order for the equation to remain true.

Base Whenever a number is raised to a power, that number is called the base.
Example: in $2^3 = 8$, 2 is the base

Basic Property An algebraic expression that is true for every possible value of the variables it contains.

Cartesian coordinate system A method of locating points on a plane. The system uses two axes that cross at right angles to each other. The axes are marked off in units starting at zero where the axes cross. One axis runs vertically, and its numbers increase as you move up the axis—this is called the y-axis. The other axis—the x-axis—runs horizontally, and its numbers increase as you move to the right.

Circumference The distance around a circle, starting at one point, going around the circle once and ending up at the same point you started.

Coefficient When a variable is multiplied by a constant, that constant is called a coefficient of the variable. Example: in the expression $4x - 2y = 12$, 4 and 2 are coefficients

Combining vectors Given two vectors A and B, the process of placing the tail of vector A at the head of vector B and then drawing a new vector starting at the tail of vector B and ending at the head of vector A. This new vector is called the resultant vector.

Commutative property of addition When you add several terms, you can change the order of addition without affecting the final answer. Example: $23 + 6 = 6 + 23$

Commutative property of multiplication When you multiply several terms, you can change the order of multiplication without affecting the final answer. Example: $23 \times 6 = 6 \times 23$

Compound interest Interest computed using both the principal and any previously accrued interest.

Constant A quantity whose value does not change, such as 5 or 12. Pi is also a constant since it always equals 3.14, rounded to the nearest hundredth.

Coordinate A number that is used to tell where a point is located on a graph. The ordered pair (7, 2) gives two coordinates that locate a unique point in the Cartesian coordinate system.

Counterexample A set of values that makes an equation false. One counterexample is all that is needed to prove an equation is not a basic property.

Cross-multiplying Multiplying the means and extremes of a proportion. If $\dfrac{3x}{4} = \dfrac{5}{12}$, then the result of cross multiplying is $4(5) = 3x(12)$.

Cube of number The third power of a number. Example: 5^3 is five cubed, or 125.

Cube root If $x^3 = y$, then x is a cube root of y.

Density The property that between any two real numbers there is another real number; the rational numbers are also dense, the integers are not.

Dependent variable A variable whose value is determined by the value of another variable in an equation.

Diameter A line segment passing through the center of a circle and connecting two points on the circle.

Difference The answer to a subtraction operation.

Direct variation A situation in which one variable is proportional to a constant multiple of another. In the equation $y = 5x$, y varies directly with x.

Distributive property The distributive property states that $a(b + c) = ab + ac$ for all values of a, b, and c. Example: $5(3 + 2) = 5 \cdot 3 + 5 \cdot 2$

Division property of equality When you divide both sides of an equation by the same nonzero quantity, the equation stays balanced.

Domain The set of all input values for a function; the allowable values of the independent variable.

Entries In a table of information, the values within the table are called entries.

Equation A mathematical sentence stating that two quantities are equal.

Expanded notation A way of writing a number so that each digit is written separately and multiplied by a power of ten written in decimal form. All the digits and their multipliers are then written as one sum.

Exponent A quantity representing the power to which a base is raised. Example: in 3^4, the exponent is 7; it means $3 \cdot 3 \cdot 3 \cdot 3$, or 343.

Exponential decay An exponential function in which either the exponent is less than one and the base is greater than one, or the exponent is greater than one and the base is between zero and one.

Exponential function An equation in which at least one variable is used as an exponent. Example: $y = 2^x$

Exponential growth An exponential function in which the exponent is greater than one and the base is greater than one.

Extremes In the proportion $\dfrac{a}{b} = \dfrac{c}{d}$, a and d are called the extremes.

Formula An equation that expresses a relationship between at least two variables.

Function A relation between two variables such as x and y where there is only one possible value of y for any given value of x.

Function notation The notation used to represent the rule between two variables, usually written $f(x)$ and read "f of x."

Greatest integer function A function that rounds input values down to the nearest integer.

Grouping symbols Parentheses or brackets used to set off one part of an equation from another.

Guess and check A method of problem solving in which a guess is made and checked in an equation. Based on the result, a new adjusted guess is made and then checked. This process is repeated until a suitable answer is found.

Horizontal line A line with slope zero. All horizontal lines are parallel to the x-axis.

Horizontal translation Moving the graph of a function to the right or left without changing the shape of the graph.

Hyperbola The graph of an inverse variation equation.

Hypotenuse In a right triangle, the side opposite the right angle; it is always the longest side.

Independent variable A variable whose value does not depend on the value of any other variable.

Integer A number with no fractional part. The numbers …–2, –1, 0, 1, 2 … represent the set of integers.

Intercept The point at which the graph of a line crosses an axis.

Interest Money charged for the use of money, usually calculated as a percentage of the amount of money borrowed.

Inverse operations Operations that undo each other, such as addition and subtraction.

Inverse variation A situation in which one variable is proportional to the inverse of another. The equation $y = \dfrac{1}{x}$ shows y varying inversely with respect to x.

Irrational number A number whose decimal representation never terminates or repeats.

Isolating a variable The process of performing operations on an equation in order to leave a chosen variable on one side of the equation and have all other expressions on the other side.

Leg Either of the two sides of a right triangle adjacent to the right angle.

Like terms Terms that contain exactly the same variables raised to the same exponents.
Example: $3x^2$ and $-2x^2$ are like terms

Linear equations An equation written in the form $y = mx + b$, where m and b are constants. The graph of a linear equation is a straight line.

Linear function A function that can be represented by a straight line not parallel to the y-axis.

Magnitude The numerical value or length of a vector.

Means In the proportion $\dfrac{a}{b} = \dfrac{c}{d}$, the means are b and c.

Metric System A system of weights and measures whose units are all based on powers of ten.

Multiplication property of equality When you multiply both sides of an equation by the same quantity, the equation remains balanced.

Multiplicative identity One is the multiplicative identity because when you multiply any quantity by one, the result is the same quantity. Example: $5 \cdot 1 = 5$

Multiplicative inverse Any two numbers whose product is one are multiplicative inverses; also called reciprocals. Example: 3 and $\dfrac{1}{3}$ are multiplicative inverses

Negative integer Any integer whose value is less than zero.

Numerical expression A mathematical expression that contains no variables.

Opposite of a vector The opposite of a vector A has the same magnitude as vector A but points in precisely the opposite direction.

Opposite of an integer Two integers are opposite if they are the same distance from zero. Example: 5 and –5 are opposites

Order of operations A set of rules determining the way to evaluate a mathematical expression. First, perform all operations within any grouping symbols. Then, perform all calculations involving exponents. Multiply or divide in order from left to right. Finally, add or subtract in order from left to right.

Ordered pair A pair of numbers enclosed by parentheses and separated by a comma. The ordered pair represents a unique point in the Cartesian coordinate plane. Example: (2, 1) is an ordered pair; it is represented by the point 2 units to the right and 1 unit up from the origin

Origin The point in the Cartesian coordinate system where the axes cross and where the values of both x and y are zero.

Parabola The graph of a quadratic equation.

Parent function The simplest form of a function that can be changed to generate an entire class of functions with similar graphs. Example: $y = x^2$ is the parent function of $y = x^2 + 2$.

Pencil test Another name for the vertical-line test.

Percent Part of 100. Example: 60% is 60 out of 100; $60\% = 0.60 = 0.6$

Perpendicular Two lines are perpendicular if they meet at right angles to each other.

Pi The ratio of a circle's circumference to its diameter.

Positive integer An integer greater than zero.

Power When you multiply the base b by itself n times, the product is the nth power of b. Example: the third power of 2 is 2^3, or 8

Power of ten If a number can be written in the form 10^n, where n is an integer, the number is a power of ten.

Power of ten notation A form where a quantity is written as a number multiplied by a power of ten.

Principal The basic amount of money used to calculate interest.

Product The result of multiplying two or more quantities.

Property of adding zero When you add zero to any quantity, the result is the same quantity. Example: $5 + 0 = 5$

Property of multiplying by one When you multiply any quantity by one, the result is the same quantity. Example: $5 \cdot 1 = 5$

Proportion property In a true proportion, the product of the means equals the product of the extremes.

Proportion A mathematical sentence stating that one ratio is equal to another. Example:

$\dfrac{a}{b} = \dfrac{c}{d}$, where $b \neq 0$ and $d \neq 0$.

Pythagorean Theorem In a right triangle, the square of the length of the hypotenuse equals the sum of the squares of the lengths of the two legs.

Quadrants The Cartesian coordinate plane is divided into four regions by the axes; these regions are called quadrants.

Quadratic equation An equation in which the highest power of the independent variable is two.

Quadratic function A quadratic equation expressed as a function. The parent function is $y = x^2$.

Quotient The result of a division operation.

Radius A line segment from the center of a circle to the circle itself.

Range (of data) The interval in which all the data lie, found by calculating the difference between the data points with the largest and the smallest values.

Range (of function) The set of all outputs for a function; the possible values of the dependent variable.

Rate of change The slope of a linear equation.

Rate of interest The ratio of the interest charged to the principal borrowed expressed as a percent.

Ratio The relative sizes of two numbers, usually expressed as $a{:}b$ or $\dfrac{a}{b}$ where $b \neq 0$.

Rational number Any number that can be written as a ratio of two integers with the second integer not equal to zero. Integers, terminating decimals, and repeating decimals are all rational numbers.

Real numbers The set of all rational and irrational numbers combined is called the set of real numbers.

Reciprocal If the product of two numbers is one, the numbers are reciprocals. A number's reciprocal is also called its multiplicative inverse.

Reciprocal function The function $y = \dfrac{1}{x}$, where $x \neq o$.

Relation Any set of ordered pairs.

Repeating decimal A decimal that after some point consists of an infinite number of repetitions of the same number sequence.

Example: $2\dfrac{2}{3} = 2.666 \ldots$

Restrictions Limitations imposed on the domain and range of a function so that all the possible inputs and outputs make sense for a real-world situation.

Resultant The one vector that represents the combination of two or more vectors.

Right angle An angle measuring exactly 90 degrees, sometimes called a square angle.

Right triangle Any triangle containing a right angle.

Rise of a line The vertical change in a line between two points, found by calculating the difference between the *y* values of the points.

Run of a line The horizontal change in a line between two points, found by calculating the difference between the *x* values of the points.

Scale factor The ratio by which the original dimensions of a figure are multiplied to enlarge or reduce the figure.

Scientific notation A shorthand way of writing very large or small numbers such that they are represented by a number between one and ten multiplied by a power of ten written in exponential form. Example: $3{,}700{,}000 = 3.7 \times 10^6$

Simple interest Interest calculated only on the original principal.

Slope The ratio of the rise of a line to its run; slope can be thought of as the "steepness" of the line.

Slope-intercept form The graph of a linear equation written in the form $y = mx + b$ has *m* as its slope and *b* as its *y*-intercept.

Square of a number A number raised to the second power. Example: 5^2 is five squared

Square root If $y = x^2$, then *x* is the square root of *y*. The square root of *a* is written $\sqrt{a}$. Example: the square root of 25, $\sqrt{25}$, is 5

Square root property The solution for $x^2 = k$, where $k > 0$ is $\sqrt{k}$ or $-\sqrt{k}$.

Standard form The standard form of a linear equation is $ax + by = c$, where *a*, *b*, and *c* are constants.

Substitute Replacing a variable with any quantity that has the same value.

Subtraction property of equality If the same quantity is subtracted from both sides of an equation, then the equation stays balanced.

Sum The result of adding two or more quantities.

Terminating decimal A decimal that stops after a certain number of digits.
Example: $\dfrac{3}{4} = 0.75$

Terms Any quantity combined as a whole with other quantities by addition.
Example: in $5x^3 - 4$, $5x^3$ and 4 are terms

Translation Moving a graph or figure up or down or to the left or right without altering its slope.

Variable A quantity, usually represented by a letter, that can take on different values.

Vectors A quantity having both magnitude and direction, often represented by an arrow.

Vertex The point at which a parabola reverses direction; also the "tip" of the absolute value function's graph.

Vertical line A line with undefined slope. All vertical lines are parallel to the *y*-axes.

Vertical-line test If *any* vertical line crosses a graph in more than one place, then the graph is not a function.

Vertical reflection An image in which each point of a graph is replaced with a point an equal distance on the other side of the *x*-axis. It looks as if the original graph has been turned upside-down.

Vertical translation Moving a graph up or down without changing the shape of the graph.

Volume The amount of space inside a three-dimensional object such as a box or a sphere.

X-axis The horizontal axis of the Cartesian coordinate system.

X-coordinate The first number in an ordered pair. It tells how far to the right or left of the origin a point is located.

X-intercept The *x*-coordinate of the point at which a line crosses the *x*-axis.

Y-axis The vertical axis of the Cartesian coordinate system.

Y-coordinate The second number in an ordered pair. It tells how far up or down from the origin a point is located.

Y-intercept The y-coordinate of the point at which a line crosses the *y*-axis.

Zero power The zero power of any number equals one. Example: $3^0 = 1$

Selected Answers

This section contains selected answers to the Practice and Problem Solving and Mixed Review sections.

Chapter 1 Integers and Vectors

Lesson 1.1, pages 4-7
Practice and Problem Solving
7. $+25$ minutes   **9.** -2340 feet
11. -8648 meters
13. $+\$36$ **15.** $+400$ square feet
17. -1000 feet
Mixed Review
19. $\$3600$

Lesson 1.2, pages 8-12
Practice and Problem Solving
7. $-8 < -6$ **9.** $-5 < -4$ **11.** $-20 < -15$
13. $-9 < 0$ **15.** 7 **17.** 0 **19.** 6 **21.** -25
23. No
Mixed Review
25. $\$487$

Lesson 1.3, pages 13-18
Practice and Problem Solving
7. 0 **9.** -15 **11.** -25 **13.** -34 **15.** -32
17. 0 **19.** -100 **21.** -3 kilowatts   **23.** $+2$
Mixed Review
25. $-86 < -28$ **27.** $0 > -86$ **29.** $+4344 > 0$

Lesson 1.4, pages 19-22
Practice and Problem Solving
3. 14 **5.** -1 **7.** 3 **9.** 32 **11.** -20
13. -18 **15.** -18 **17.** $\$897.66$ **19.** $\$479.53$
Mixed Review
21. 6.06 inches   **23.** 2.85 inches
25. 4.91 inches   **27.** -123 feet   **29.** $+118$ feet

Lesson 1.5, pages 24-28
Practice and Problem Solving
7. 2 **9.** -9 **11.** -7 **13.** -72 **15.** $8\dfrac{1}{3}$

17. $-4\dfrac{3}{4}$ **19.** $-2; 3; -1; -3; 3$
21. $17; -10; -11; 21; -17$
Mixed Review
23. 11 **25.** $\$40$
27. Copying 500 sheets costs less.

Lesson 1.6, pages 29-33
Practice and Problem Solving
5. -2 **7.** 11 **9.** 7 **11.** -2 **13.** -53
15. 55 **17.** -85 **19.** -66
Mixed Review
21. $\$40$ **23.** 5217.625 **25.** 5203.5

Lesson 1.7, pages 34-39
Practice and Problem Solving
7.

9. Air speed is greater.
11. Air speed is less.
Mixed Review
13. $-\$360; +\$141; +\$317; -\$2$ **15.** $10; -1;$
$-4; -10; -16; 8; 0; -3$ **17.** 38 **19.** 83

Chapter 2 Scientific Notation

Lesson 2.1, pages 4-8
Practice and Problem Solving
5. 3^2 **7.** $(-5)^2$ **9.** $(-3)^5$ **11.** 27 **13.** -1
15. 128 **17.** 1,000,000 watts **19.** 50,000
21. 3^4 **23.** 4^5
Mixed Review
25. Cathy, Tim, Jose, Maria, Pat **27.** -1 pound
29. -505 feet **31.** -475 feet

Lesson 2.2, pages 9-12
Practice and Problem Solving
7. 0.00412 **9.** 0.1 **11.** 0.04 **13.** 1 **15.** 0.001

17. 3^{-4} **19.** 10^{-2} **21.** $\dfrac{1}{1000000000}$;

0.000000001
Mixed Practice
23. Volga < Mississippi
25. Amazon $+2100$ miles; Danube -124 miles;
Ganges -340 miles; Mississippi $+440$ miles;
Nile $+2260$ miles; Ohio -590 miles;
St. Lawrence -1100 miles; Volga $+294$ miles

Lesson 2.3, pages 13-15
Practice and Problem Solving
5. 93×10^6 **7.** 55×10^8
9. 25×10^{-6} millimeters **11.** $\$4 \times 10^{12}$
13. $2 \times 10^4 + 3 \times 10^3 + 6 \times 10^2 + 8 \times 10^1 +$
1×10^0
15. $7 \times 10^2 + 2 \times 10^1 + 6 \times 10^0 + 0 \times 10^{-1} +$
5×10^{-2}
Mixed Review
17. -7 **19.** -9 **21.** -19 **23.** $\$0.35$

Lesson 2.4, pages 16-21
Practice and Problem Solving
5. 5.67×10^1 **7.** 2.45×10^5 **9.** $9.3673 \times$
10^4
11. 4.5×10^7 **13.** 1.35737×10^2
15. 2.0×10^{19} molecules **17.** 6.5×10^{-6}

19. 0.00001 **21.** 5,870,000,000,000 miles
23. 0.0000003 meters
Mixed Review
25. 6 months **27.** 105 seconds

Lesson 2.5, pages 22-27
Practice and Problem Solving
7. 10^2 **9.** 10^{-3} **11.** 10^{-13} **13.** 10^0
15. 3.5×10^9 **17.** 3.18×10^{-8}
19. 18×10^{-1} meters; 1.8×10^{-2} meters;
1.8 meters
Mixed Review
21. $36°F$ **23.** $6 \times 10^3°C$

Lesson 2.6, pages 28-32
Practice and Problem Solving
5. 0.034 kilometers **7.** 58 millimeters
9. 0.350 seconds **11.** 800,000 milligrams
13. 0.083 meters **15.** 0.142 amps **17.** 10^{-12}
Mixed Review
19. $-10°F$ **21.** $-15°F$
23. 1.392×10^6 kilometers **25.** 5.4×10^{-15}

Chapter 3 Using Formulas

Lesson 3.1, pages 4-8
Practice and Problem Solving
5. 65 **7.** 7 **9.** 2 **11.** -28 **13.** -100
15. 20 **17.** -22 **19.** 48 **21.** 4 **23.** 3
25. $\$106$
Mixed Review
27. 4,700,000,000 years **29.** 3×10^{-2}

Lesson 3.2, pages 9-16
Practice and Problem Solving
5. yes **7.** yes **9.** no **11.** yes **13.** yes
15. 384 square millimeters **17.** 12 miles
19. 81 square feet **21.** $95°F$ **23.** same

25. $D = RT$ **27.** $T = \dfrac{D}{R}$ **29.** 15 miles

Mixed Review

31. Charlotte, Long Beach, Austin, New Orleans, Cleveland, Seattle

33. 2.76×10^{-8} centimeters **35.** 0.048 meters

Lesson 3.3, pages 17–20

Practice and Problem Solving

5. 12.57 meters; 12.57 square meters **7.** 50.27 centimeters; 201.06 square centimeters

9. 39.27 millimeters; 124.69 square millimeters

11. 38.48 square centimeters

13. 25.13 square meters **15.** 157.08 square feet

Mixed Review

17. Pacific Ocean: −12,925 feet; Atlantic Ocean: −11,730 feet **19.** 4.03×10^2 inches

Lesson 3.4, pages 21-24

Practice and Problem Solving

5. 15.625 cubic feet

7. about 113.1 cubic meters

9. about 2.61×10^{11} cubic miles

11. about 268.1 cubic centimeters

13. 3456 cubic inches

Mixed Review

15. On–Time **17.** −$295; $1655; −$695

19. +$665

Lesson 3.5, pages 25-28

Practice and Problem Solving

3. $3.60 **5.** $90.77 **7.** $1027.57

9. $2382.02

Mixed Review

11. Line B: $105,000 **13.** The order from left to right is B, C, D, A. **15.** 4.167×10^0

17. $135 **19.** 14,400 pounds

Chapter 4 Solving Equations

Lesson 4.1, pages 4-11

Practice and Problem Solving 4-11

7. −8 **9.** 8 **11.** $6\frac{2}{3}$ **13.** $-9\frac{2}{3}$ **15.** $-13\frac{1}{3}$

17. 3 **19.** $\frac{1}{2}$ **21.** −30 **23.** $\frac{V}{r} = i$

25. $\frac{i}{rt} = p$ **27.** 15 feet

Mixed Review

29. 2400 meters **31.** 0.52 liters

Lesson 4.2, pages 12-16

Practice and Problem Solving

5. $\frac{1}{4}$ **7.** $1\frac{1}{9}$ **9.** −84 **11.** 25 **13.** $7\frac{1}{3}$

15. −25 **17.** $A = LW$

Mixed Review

19. 5.6×10^{-5} **21.** 1.4395×10^2 **23.** The change in the temperature is −10°. **25.** 1 gallon of paint costs $8.50.

Lesson 4.3, pages 17-22

Practice and Problem Solving

5. 4 **7.** −40 **9.** 30 **11.** 25% **13.** 66.7%

15. 4.5% **17.** $0.60 **19.** 25% **21.** $54.25

23. $3000

Mixed Review

25. 2.478×10^{-4} **27.** −37

Lesson 4.4, pages 23-29

Practice and Problem Solving

5. 4 **7.** 63 **9.** −1 **11.** −2 **13.** −0.8

15. 3.2 **17.** 4.4 **19.** −12.7 **21.** −3.12

23. 24 workstations

Mixed Review

25. $73 **27.** 25 = yard line

29. 163 miles each day.

Lesson 4.5, pages 30-35

Practice and Problem Solving

3. 7 **5.** −6 **7.** −6 **9.** −12 **11.** −1.78

13. −2 **15.** $x = \dfrac{c - b}{a}$ **17.** 54 feet

19. 11 feet **21.** $\dfrac{P - 2W}{2} = L$ **23.** 96 **25.** $125

Mixed Review

27. 1.3312×10^3 **29.** about 9 centimeters

Lesson 4.6. pages 36-40

Practice and Problem Solving

5. $\dfrac{6}{7}$ **7.** $-3\dfrac{4}{7}$ **9.** $6 \neq 5$; no solution   **11.** -8

13. $-2\dfrac{6}{7}$ **15.** $\dfrac{6}{41}$ **17.** $60 + x$ **19** 30 volts;
90 volts.   **21.** $1.5(2\pi x) = 2\pi(x + 4)$ **23.** 70
42-inch; 280 38-inch   **25.** 10 by 4; 8 by 6.
27. $-126, +111, -7, -40, +32$

Chapter 5 Graphing Linear Functions

Lesson 5.1, pages 4-10

Practice and Problem Solving

5. $A(-4,0)$; $B(2,-3)$; $C(1,5)$; $D(-2,-2)$
7. Caddo $(-6,7)$; Limestone Gap $(-4,0)$;
Ada $(4,1)$; Windom $(-6,-3)$ Hartshorne
$(5,-4)$; Bug Tussle $(0,-6)$ **9.** four   **11.** two

Mixed Review

13. true   **15.** 10^{-5} kilometers   **17.** about
132,889.37 centimeters   **19.** 120 votes
21. 50%   **23.** 1

Lesson 5.2, pages 11-16

Practice and Problem Solving

7. rectangle   **17.** straight line

19.

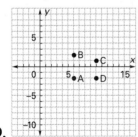

Mixed Review

21. 1.3986×10^7 **23.** 14   **25.** \$351
27. -10 **29.** -12 **31.** 166.7

Lesson 5.3, pages 17-23

Practice and Problem Solving

7. $-\dfrac{1}{2}$ **9.** 5   **11.** $\dfrac{1}{5}$ **13.** $-\dfrac{4}{3}$ **15.** 2

17. $-\dfrac{1}{11}$ **19.** 0   **21.** 2.3 feet per mile

23. $\dfrac{1}{10}$

Mixed Review

25. 44.5 gallons per hour   **27.** 30 feet   **29.** 24

Lesson 5.4, pages 24-32

Practice and Problem Solving

7. slope $= -3$; y–intercept 5   **9.** slope $= -1$;
y-intercept 8   **11.** slope $= \dfrac{5}{2}$; y-intercept $\dfrac{3}{2}$

13. slope $= \dfrac{5}{2}$; y-intercept $\dfrac{3}{2}$ **15.** 20; 10

17. 6   **19.** 0.05; 2
Mixed Review
21. \$8333.33   **23.** 7 weeks

Lesson 5.5, pages 33-37

Practice and Problem Solving

5. $\dfrac{5}{3}$; 5   **7.** 8; -4 **9.** $\dfrac{5}{3}$; -2.5 **11.** $-\dfrac{5}{2}$; 10

13. 6; -4 **15.** -8; $\dfrac{8}{3}$ **17.** $d = 4t + 2$

19. $A = 6 + 8h$
Mixed Review
21. 10^{-5} **23.** 10^8 **25.** 306,796.16 square
centimeters   **27.** 31

Lesson 5.6, pages 38-44

Practice and Problem Solving

5. $y = x + 1$ **7.** $y = \dfrac{3}{2}x + \dfrac{19}{2}$ **9.** $y = x + 5$

11. 65 miles per hour **13.** $5x + 3y = 65$
15. 10 adult tickets and 5 child's tickets
17. $d = \dfrac{1}{500}F$ **19.** 1500 pounds
Mixed Review
21. –12 pounds **23.** –9° **25.** 200 **27.** 120 casings

Lesson 5.7, pages 45-47
Practice and Problem Solving
7. They are equal. **9.** They are perpendicular.
Mixed Review
13. 8 **15.** –33 **17.** 1 **19.** $y = 3x - 2$
21. $2,743.75 **23.** $2,100,000

Chapter 6 Nonlinear Functions

Lesson 6.1, pages 4-9
Practice and Problem Solving
7. $y = 2x - 2$; 8, 10, 12 **9.** $y = 5x + 1$; 26, 31, 36
11. $y = \dfrac{1}{2}x + \dfrac{1}{2}$; 3, $\dfrac{7}{2}$, 4 **13.** –7 **15.** 2 **17.** 0
19. $y = 8x + 15$; slope 8; y-intercept 15
Mixed Review
21. 8.5×10^{-5} **23.** $5\dfrac{4}{5}$ **25.** –2 **27.** 5 **29.** 61.5%

Lesson 6.2, pages 10-17
Practice and Problem Solving
7. vertical stretch by a factor of 2 **9.** reflection across the x-axis, vertical move up 1 unit, vertical stretch by a factor of 3 **11.** stretch by a factor of 4, vertical move up 6 units **13.** stretch by a factor of 3, vertical move up 2.5 units
Mixed Review
15. $216.08 **17.** 10^{-5} **19.** 45 **21.** 1600 **23.** 20
25. 20°; 60°; 100° **27.** $-\dfrac{3}{2}$; 6

Lesson 6.3, pages 18-24
Practice and Problem Solving
7. 100 watts **9.** 1600 watts **11.** –4 **13.** –10

15. vertical move 1 unit down **17.** horizontal move 2 units left, vertical move 1 unit up, vertical stretch by a factor of 3 **19.** vertical stretch by a factor of π
Mixed Review
21. 4AM **23.** –7°F **25.** 18,75% **27.** $3\dfrac{3}{4}$

29.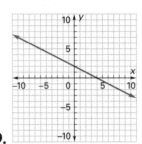

Lesson 6.4, pages 25-32
Practice and Problem Solving
7. –50 **9.** about 24.8 **11.** 4.90 and –4.90
13. 5 meters **15.** 5 feet **17.** 229.04 inches
19. about 10.63 miles **21.** 91 feet
Mixed Review
23. –9 **25.** –4; –9 **27.** about 34.3 gallons
29. $311.52

Lesson 6.5, pages 33-38
Practice and Problem Solving

7.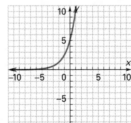

9.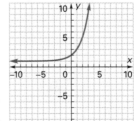

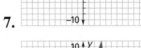

11.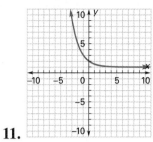

13. Multiply by a negative number.

15. $\dfrac{1}{2}, \dfrac{1}{4}, \dfrac{1}{8}, \dfrac{1}{16}$

17. $b = p(1.06)^x$ $1338.22 **19.** $14,327.86

Mixed Review

21. $4166.67 **23.** –6 **25.** $y = -\dfrac{2}{3}x + \dfrac{11}{3}$

Lesson 6.6, pages 39-44

Practice and Problem Solving

9. The graph moves away from the origin.
13. The graph moves away from the origin and reflects over the x-axis. **17.** The graph moves closer to the x- and y-axes. **19.** all the integers
21. The graph looks like a staircase.
23. 3 stamps; You can only buy a whole number of stamps.

Mixed Practice

25. under **27.** 5 **29.** about 16.97 feet
31. 19 inches

Index

Credits

Algebra at Work

ix: top, Tom Carroll/Photo, Inc.; bottom, © 1996 PhotoDisc, Inc.x: B. Daemmrich/The Image Works. xvi: top, L. Kolvoord/The Image Works; bottom, David Young-Wolff/Tony Stone Images.

Chapter 1

1: Bob Kramer/Picture Cube. 2: Neil Nissing/FPG. 4: Mark Burnett/Stock Boston. 7: ©1995 PhotoDisc, Inc. 8: Jack Plekan/FundamentalPhotographs. 12: ©1995 PhotoDisc, Inc. 13: ©1996 PhotoDisc, Inc. 15: ©1996 PhotoDisc, Inc. 17: Peter Menzel/Stock Boston. 21: David Frazier/Photo Researchers, Inc. 22: John Coletti/The Picture Cube. 28: Tom Pantages. 29: Wernher Krutein/Liaison International. 32: COREL. 38: © 1995 PhotoDisc, Inc. 40: CORD. 41: CORD. 45: COREL. 49: Nigel Catlin/Photo Researchers, Inc. 56: David Weintraub/Stock Boston. 57: Seth Resnick/Stock Boston. 61: © 1995 Photo Disc, Inc. 63: COREL.

Chapter 2

1: © 1995 PhotoDisc, Inc. 3: Michael Rosenfeld/Tony Stone Images. 9: Bill Gallery/Stock Boston. 12: top, Courtesy of International Business Machines Corporation. Unauthorized use not permitted; bottom, © 1996 PhotoDisc, Inc. 13: Will & Deni McIntyre/Tony Stone Images. 17: Robert Reichert/Liaison International. 19: Michael Tamborrino, The Stock Shop. 20: Seth Resnick/Liaison International. 24: Charles Gupton/Stock Boston. 25: © 1995 PhotoDisc, Inc. 26: James D. Wilson/Gamma Liaison. 28: Ed Lallo/Liaison International. 33: Stephen Frisch/Stock Boston. 34: CORD. 37: John David Fleck/Liaison Internaitonal. 38: Armen Kachaturian/Liaison International. 41: © 1995 PhotoDisc, Inc. 44: Ben Simmons/The Stock Market.

Chapter 3

1: © 1995 PhotoDisc, Inc. 3: Michael Newman/PhotoEdit. 4: Cathlyn Melloan/Tony Stone Images. 5: D&I MacDonald/The Picture Cube. 9: Bill Gallery/Stock Boston. 12: Bob Daemmrich/Tony Stone Images. 13: Paul Silverman/Fundamental Photographs. 15: Bonnie Kamin/PhotoEdit. 17: left, Nathan Benn/Stock Boston; right, Tom Bean/Tony Stone Images. 18: © 1995 PhotoDisc, Inc. 20: left, Jeffrey Reed/The Stock Shop; right, Roger Tully/Tony Stone Images. 21: © 1995 PhotoDisc, Inc. 22: Tom Pantages. 24: © 1995 PhotoDisc, Inc. 28: Glen Allison/Tony Stone Images. 29: CORD. 32: CORD. 35: Brian Seed/Tony Stone Images. 37: Peter Vandermark/Stock Boston. 39: Courtesy of Federal Express Corporation. 41: Bruce Ayers/Tony Stone Images. 42: Courtesy of Aetna Life & Casualty Company. 43: Paula Lerner/The Picture Cube. 45: © 1995 PhotoDisc, Inc.

Chapter 4

1: © 1996 PhotoDisc, Inc. 3: Dick Luria/The Stock Shop. 4: Phil Matt. 6: John Lei/Stock Boston. 8: James Schnepf/Liaison International. 12: Courtesy of Toyota Manufacturing USA, Inc. 14: David Dewhurst.16: © 1995 PhotoDisc, Inc. 17: © 1995 PhotoDisc, Inc. 20: Lawrence Migdale/Photo Researchers, Inc. 22: Tom Pantages. 23: Courtesy of RJR Nabisco, Inc. 28: Lester Sloan/Liaison International. 29: © 1996 PhotoDisc, Inc. 30: © 1995 PhotoDisc, Inc. 33: Seth Resnick/Stock Boston. 34: Focus on Sports. 36: James Schnepf/Liaison International. 38: © 1996 PhotoDisc, Inc. 40: Tom Tracy/The Stock Shop. 41: © 1995 PhotoDisc, Inc. 46: CORD 48: CORD. 53: Bill Bachman/PhotoEdit. 56: Scott Barrow. 60: F. Pedrick/The Picture Cube, Inc. 62: Buenos Dias Bildagentur/Liaison International.

Chapter 5

1: Courtesy of New York Stock Exchange. 4: Peter LeGrand/Tony Stone Images. 7: Jeff Greenberg/The Picture Cube, Inc. 10: Cindy Charles/PhotoEdit. 11: Richard Pasley/Stock Boston. 12: Courtesy of Champion Spark Plug Company. 14: © 1995 PhotoDisc, Inc. 15: © 1995 PhotoDisc, Inc. 17: Mark E. Gibson. 20: Kenneth Gabrielsen/Liaison International. 23: Jeff Greenberg/PhotoEdit. 24: © 1996 PhotoDisc, Inc. 28: Mary Kate Denny/PhotoEdit. 30: Stephen Frisch/Stock Boston. 31: George Goodwin/The Picture Cube, Inc. 33: Bob Daemmrich/Stock Boston. 35: © 1995 PhotoDisc, Inc. 40: © 1995 PhotoDisc, Inc. 42: © 1995 PhotoDisc, Inc. 43: J. Froscher/The Image Works. 48: CORD. 49: CORD. 54: left, © 1995 PhotoDisc, Inc.; right, © 1996 PhotoDisc, Inc. 56: Zoological Society of San Diego. 60: © 1995 PhotoDisc, Inc. 65: Rob Crandall/Stock Boston.

Chapter 6

1: Gabe Palmer/Mug Shots. 2: The News Tribune, Tacoma, WA, 1940. 4: © 1995 PhotoDisc, Inc. 6: George Hunter/Tony Stone Images. 14: Charles D. Winters/Photo Researchers, Inc. 15: © 1995 PhotoDisc, Inc. 17: Lincoln Russell/Stock Boston. 18: top, © 1995 PhotoDisc, Inc.; bottom, Peter Menzel/Stock Boston. 20: NASA/Photo Researchers, Inc. 23: Ray Ellis/Photo Researchers, Inc. 24: Eglin Air Force Base. 27: © 1995 PhotoDisc, Inc. 29: Chester Higgins/Photo Researchers, Inc. 31: Catherine Noren/Photo Researchers, Inc. 33: Professor Motta/Science Library/Photo Researchers, Inc. 35: Wesley Bocxe/Photo Researchers, Inc. 38: Harrison Baker, Ottawa, Canada. 41: Donald Dietz/Stock Boston. 44: Michael Schwartz/Liaison International. 46: CORD. 47: CORD. 50: top, © 1995 PhotoDisc, Inc.; bottom, Steve Starr/Stock Boston. 51: © 1995 PhotoDisc, Inc. 53: Stacy Pick/Stock Boston. 54: Robert Goldstein/Medichrome. 59: Greenlar/The Image Works. 62: Mark Gibson. 64: Thomas Fletcher/Stock Boston.